Fine Homebuilding

TRICKS OF THE TRADES:

BUILDING METHODS AND MATERIALS

D0001594

The Taunton Press

Cover Illustration: Robert La Pointe

for fellow enthusiasts

© 1994 by The Taunton Press, Inc.

First printing: March 1994
Printed in the United States of America

A Fine Homebuilding Book
Fine Homebuilding® is a trademark of The Taunton Press, Inc.,
registered in the U.S. Patent and Trademark Office.

The Taunton Press, 63 South Main Street, Box 5506,
Newtown, CT 06470-5506

Library of Congress Cataloging-in-Publication Data

Tricks of the trades. Building methods and materials.
            p.   cm.
    At head of title: Fine homebuilding
    "A Fine homebuilding book"—T.p. verso.
    Includes index.
    ISBN 1-56158-077-5
    1. Building — Miscellanea.   2. Building materials — Miscellanea.
I. Fine homebuilding.   II. Title: Building methods and materials.
    TH153.T724   1994   93-48047
    690'.83 — dc20   CIP

CONTENTS

INTRODUCTION

The ingenuity of people who work with their hands is staggering. They keep coming up with better ideas on time-tested techniques and improved uses for well-known materials. We who build professionally or do much of our own work around our houses are appreciative beneficiaries of these ideas, which serve us well in our work.

The best gifts provide lasting pleasure, and this book is a collection of such gifts. Here are the site-proven tips of ingenious people creating imaginative solutions to common problems. Like its companion books, this volume represents the first time these items, collected from *Fine Homebuilding* magazine's most popular columns, have appeared in book form.

The two magazine columns have similar benefits but different methods. Tips & Techniques is a forum for the ideas that people want to share with their fellow builders and carpenters. Q & A takes serious questions from people who face a problem they can't solve satisfactorily and provides answers from top-notch professionals who have faced the problem before. Both columns are avidly followed by readers of *Fine Homebuilding*, many of whom read these columns before they read anything else in an issue. Indeed, quite a few readers claim these columns alone are worth the price of a subscription.

The focus of this book is techniques that will improve your building methods and your uses of building materials. Here are tips that are sure to serve both the do-it-yourselfer and the professional carpenter the rest of their lives—or until another ingenious person comes along with a further improvement. That's likely to be printed in a future issue of *Fine Homebuilding*, too.

1

FRAME CARPENTRY

QUICK-RELEASE TRUSS ANCHOR

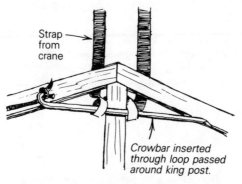

Strap from crane

Crowbar inserted through loop passed around king post.

The house that we worked on this spring has a roof built with scissor trusses. We set them with a crane, but instead of using a chain to hoist them and a system of scaffold planks to gain access to the chain to release it, we devised the nylon-strap and crowbar-pin setup shown above. Looped in this fashion, the nylon strap holds the crowbar in place by tension. After a truss is positioned, nailed-off and braced, the crane operator loosens the strap and an assistant on the deck yanks a rope tied to the crowbar, releasing the strap from the truss. Yes, the crowbar does fall with some force, and anybody in the vicinity should be wearing a hardhat.

—Mike Nathan, Stella, N. C.

TRUSS-ANCHOR REFINEMENT

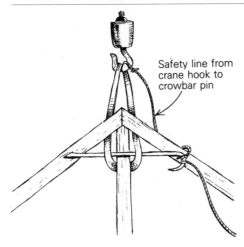

Safety line from
crane hook to
crowbar pin

Here's an addition that I think improves upon Mike Nathan's tip for
lifting trusses (see the facing page). I realize that Mr. Nathan and his
crew protect themselves with hardhats, but a safety line from the
crane's hook to the crowbar pin (as shown in the drawing above)
will keep the crowbar from falling after the truss has been set and
the crowbar has been pulled. This will protect bystanders without
proper headgear and will prevent possible damage to roughed-in
gas or water lines.

—Eugene F. Oakley, El Sobrante, Calif.

DRESSING BIG BEAMS

The truck was already overdue when it pulled up with its load of
roughsawn 6x12 timbers. Even though we'd ordered surfaced
material, we weren't going to refuse delivery; we'd just have to deal
with it. We thought about using our 7-in. grinder and the little 3-in.
power planer that we had on site, but figured that it would take
forever to dress the beams with these tools, and that we'd likely
trash them in the process.

Solution: we rented a floor sander. Using various grits of paper,
we walked it back and forth on the timbers. It did the job in almost
no time, and we saved our other tools for minor touch-ups.

—Greg Braun, Cabool, Mo.

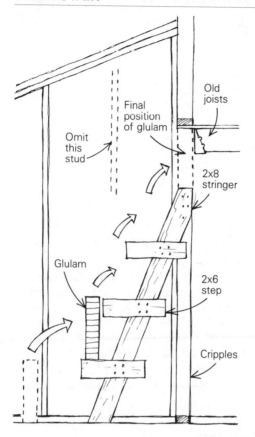

In the course of adding a large family room to a house, we had to place a 22-ft. glulam atop an old stud wall to carry the weight of a couple of rooms above. The glulam weighed almost 500 lb., and we had no access for a crane, boom-truck or a forklift. After easing the beast into the house with rollers and a temporary ramp that went from the driveway to the first floor, we devised what we called a "beam-stair" for making short lifts, one end at a time.

Once the ceiling joists were temporarily shored up, we removed the old wall and nailed in three cripples on each side. As we framed the adjacent walls, we left out a stud on each side for maneuvering room. In front of the double-width stud spaces, we built two temporary stair-step arrangements, dividing the height of the lift into four intervals of about 22 in. each. As shown above, the stringer was a 2x8 affixed to the cripples with duplex nails and tacked securely at the bottom to the subfloor. The "steps" were 2x6 blocks, each about 18 in. long, secured to the stringer with four duplex nails apiece. They were canted toward the stringer a bit, so the beam would tend to tilt a little backward toward the stringer rather than forward onto us.

When the beam-stairs on each side were ready, we gathered three crew members plus the owner, and lifted one end of the beam at a time onto a step, and so on, until we had the beam sitting on the cripples. One end of the beam was always supported, so we didn't have to stagger with it free in the air. After a little more jacking, prying and shimming, we nailed it in place, added joist hangers for the second-floor joists we had previously cut, and took all the temporary stuff away.

—Robert Gay, Seattle, Wash.

JOIST LIFT

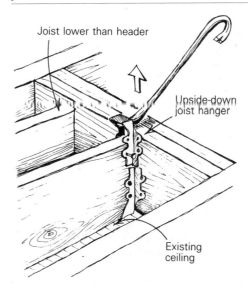

Joist lower than header

Upside-down joist hanger

Existing ceiling

During a recent remodeling project, we had to put a new floor over an old ceiling. The cramped quarters made it impossible to swing a hammer from below to bring the joists flush with newly added headers and beams. The drawing above shows how we used a temporary upside-down joist hanger and a crowbar to bring a joist flush, allowing its hanger to be installed in the right position.

—Jim Lockwood, Brookline, Mass.

GUSSET BLOCKS

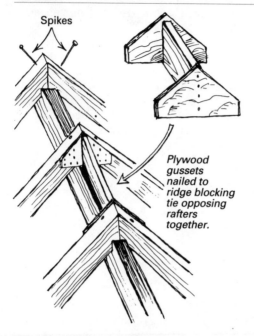

Spikes

Plywood gussets nailed to ridge blocking tie opposing rafters together.

When I built my backyard guest cottage, I used long spikes to affix the rafters to the ridge beam, and plywood gussets to tie the opposing rafters together. The rafters, some of which had nasty twists in them, also needed to be blocked at the top and bottom to keep them straight. Always looking for a way to save a little time, I prefabricated some combination gusset/blocks out of 2x and ½-in. plywood scraps (shown above). The length of the blocks equals the distance between the rafters, minus an inch to allow for two thicknesses of plywood. I used my pneumatic nailer to assemble the parts and nail them home. The system proved to be quick and sturdy. And instead of the typical steel-strap detail over the tops of the rafters, this method left the tops of the rafters clear for attaching the roof sheathing.

—Ron Milner, Grass Valley, Calif.

NON-SKID SCAFFOLD

I build in the Pacific Northwest, which is the same as saying I build in the rain. The wet weather makes for slippery footing, so to get a better grip on the ground I tack leftover asphalt shingles to the planks, ramps and scaffolds that surround our job sites. In my experience, the shingles are unaffected by rain and provide better traction than wooden cleats.

—Darrell Ohs, Nanaimo, B. C.

STRAIGHTENING CROOKED LUMBER

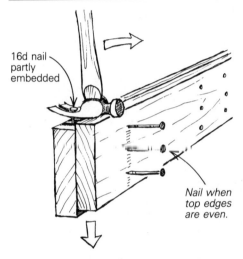

16d nail partly embedded

Nail when top edges are even.

Building built-up girders, headers and beams can be a nuisance if the components aren't straight. The drawing above shows a simple way to use a hammer and a 16d nail to lever a pair of diverging sticks back into line. Drive the nail as far as possible into the low piece, leaving enough of its shank exposed to get the hammer claws under its head. Then use the high piece as the fulcrum as you start to withdraw the nail. The high piece should come into alignment, allowing you to spike the pieces together properly. You can also use this method to align wayward joists to their rim joist.

—Jeff Fosbre, Dunellen, N. J.

STRAIGHTENING STUDS

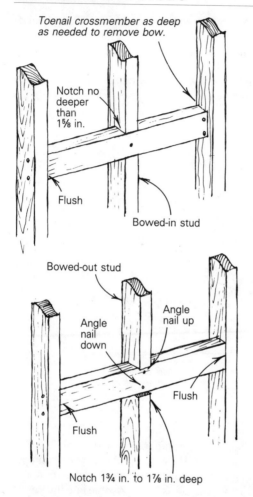

Toenail crossmember as deep as needed to remove bow.

Notch no deeper than 1⅝ in.

Flush

Bowed-in stud

Bowed-out stud

Angle nail up

Angle nail down

Flush

Flush

Notch 1¾ in. to 1⅞ in. deep

It's not unusual for a stud wall to have some sticks in it that bow in or out of the wall plane, making it tough to do a decent job on the drywall or paneling. Here's how I fix both situations. After identifying the bowed-in studs, I work on the worst one first. Using a straightedge held vertically against the side of the stud, I find the high point of the bow and measure across it for a notch that will accept a 2x4 crossmember. Then I set the saw depth to make a cut just a little deeper than the thickness of my 2x4 crossmember—about 1⅝ in. Now I nail the crossmember to the bowed stud, flush one end of it to the adjacent stud and nail it, and toenail the opposite end as deep as needed to remove the bow (top drawing, above).

To fix a bowed-out stud (bottom drawing on facing page), I make a notch 1¾ in. to 1⅞ in. deep at the point of the stud that is bowed out the most. Then I drive a pair of 16d nails—one angled up and one angled down—to anchor the crossmember to the stud. When the crossmember is flushed and nailed with the adjacent studs, the bow is gone. This second method works well when the opposite side of the wall is inaccesible—covered with a shear wall, for example.

—John Riedhart, Ventura, Calif.

MORE ON STRAIGHTENING STUDS

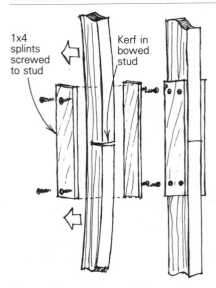

1x4 splints screwed to stud

Kerf in bowed stud

Here's a variation on John Riedhart's methods (on the facing page) for straightening a bowed stud. As shown in the drawing above, make a kerf in the edge of the stud at the apex of the bend. Use a thin blade for this so the stud will be able to close on itself when pressure is applied. As you apply that pressure, screw a pair of 1x4 splints to the sides of the stud. This way you can easily adjust the splints if you don't get it right the first time. The splints will restore the stud's rigidity and replace the compressive strength lost by a sawcut that doesn't completely close on itself.

—Tore J. Wubbenhorst, Lindenhurst, N. Y.

THE TWISTER

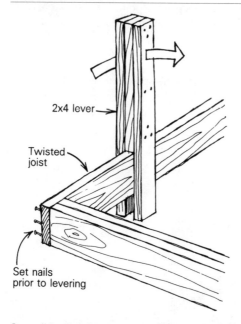

One of the first things we do on a new framing site is to build a "twister" to help straighten corkscrewed lumber. Our twisters are made of two 3-ft. 2x4s and one 2-ft. 2x4. As shown here, the long 2x4s sandwich the short one, creating a slot at one end. To use the twister, we nail the twisted piece of stock at one end. Then we slip the tool over the other end and move the twister until the stock comes flush with its nailing surface. The twister usually provides enough leverage that it only takes one hand, leaving the other free to swing a hammer or fire a nailer.

—Sean Sheehan, Basin, Mont.

FLUSH FRAMING MEMBERS

Framing members brought flush with pipe clamp.

1½ -in. by 3-in. by ⅜-in. steel plate

I use a modified pipe clamp to bring adjacent framing members into alignment. As shown above, I welded a ⅜-in. steel plate to the stationary head and the sliding foot on one of my clamps. The larger bearing surface makes it easy to bring studs, top plates, trimmers and other framing materials that may be warped into the same plane before nailing them together.

—Mike Nathan, Hailey, Idaho

RETROFIT JOISTS

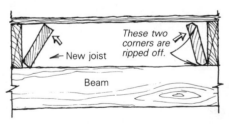

New joist

These two corners are ripped off.

Beam

If you've ever had to beef up a pier-and-beam floor with some additional joists tucked between the subfloor and the beams, you know it can be difficult work in a space where it's tough to maneuver tools and materials. The diagonal dimension of the joist is too big to let it roll into place without some serious pounding. After pondering the problem for a while, I got an idea. If I set my saw at 45° and ripped away the top edge on one side of the joist and the bottom edge on the other, as shown above, the joist should roll a lot easier and I would still have my original dimension. It works.

—Don Boerner, Rowlett, Tex.

ADDING NEW JOISTS

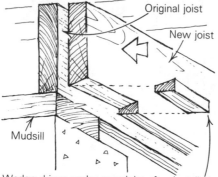

Original joist

New joist

Mudsill

Wedge driven under new joist after installation

The drawing above shows how I strengthened the original joists in my house by sistering on some new ones. The typical problem encountered during this operation is trying to get the ends of the joists to fit between the mudsills and the subfloor. When tilted, the joist is a bit too big to fit without serious persuasion, usually administered from a contorted position. To avoid this, I cut a wedge from each end of a new joist, allowing me to tip the joist into position easily. Then I tapped in the wedges and nailed them in place for full bearing.

—*Roger Westerberg, Verndale, Minn.*

STRENGTHENING OLD JOISTS

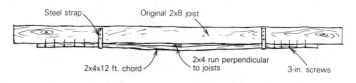

Steel strap

Original 2x8 joist

2x4x12 ft. chord

2x4 run perpendicular to joists

3-in. screws

Anyone who has ever renovated an old house has probably dealt with a floor system that is a bit too springy. It seems that 2x8 rough-cut joists were once used everywhere, regardless of the span.

I recently started work on a home built in the 1890s that has 2x8 joists on 16-in. centers, spanning 14 ft. They were also installed without blocking and were beginning to twist in spots. The basement has a dirt floor but plenty of headroom and usable floor space, so I didn't want to add a midspan beam with columns. Beefing up the joists was the only answer.

For blocking, I ran a 2x4 perpendicular to the joists across their bottoms at the center of the span. With the butt end of the 2x4 against the foundation, I worked my way across the floor, plumbing the joists and screwing through the 2x4 into each joist to secure them.

Next I turned each joist into a bowstring truss. I used drywall screws, steel straps and construction adhesive to attach 2x4x12-ft. chords to the bottom of each joist, as shown at the bottom of the facing page. I used five 3-in. drywall screws per side to pull the 2x4s into place over the block in the center of each joist. Then I added a pair of steel straps to get some fasteners working in shear in addition to withdrawal. This took only a few hours to do, and the floor now feels like it has 2x12 joists.

—*David Wallace, Annapolis, Md.*

BEEFING UP OLD JOISTS

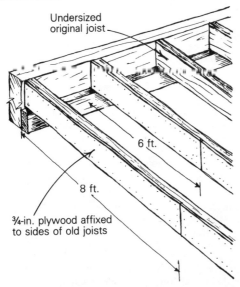

Undersized original joist

6 ft.

8 ft.

¾-in. plywood affixed to sides of old joists

I've been restoring a 120-year-old farmhouse that had 14-ft. 3x8 joists made of hemlock. Most of them were still sound, but their long, unsupported spans were pretty springy. My client didn't like the bouncy feel of the floor, and I was concerned that the flex might cause some of the seams in the drywall to crack in the partition walls. Given the minimal clearance under the house and the excavation necessary to dig footings for a midspan beam, we decided to strengthen the existing joists. But with what? If we sistered a 2x8 onto each old joist, the lumber alone would cost about $1,300. As an alternative, we asked our lumberyard about the cost of ripping enough ¾-in. plywood to serve the same purpose. Price for plywood plus cutting charge: $600.

As shown in the drawing on p. 19, we ended up going with the plywood method. We staggered the joints, and scabbed the plywood splints onto the sides of the old joists using liberal amounts of construction adhesive and 8d nails, driven with a nail gun. As we applied the adhesive to the joists, we also ran a bead of it along the top edge of the plywood. This helped to affix the subfloor to the plywood, and substantially cut down on floor squeaks. The result is a very stiff floor system.

—Richard E. Reed, Doylestown, Pa.

FLOOR GUARD

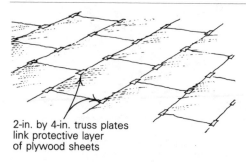

2-in. by 4-in. truss plates
link protective layer
of plywood sheets

As we began a recent remodeling project, our first task was to protect the terrazzo floor. We put down a layer of black plastic, followed by a layer of ½-plywood. The problem was, how to keep the plywood sheets linked tightly together throughout the duration of an extensive remodel. We knew duct tape would be tedious to install and it would likely pull up or get torn in places. Then my colleague Dennis Johnson had an idea—truss plates. We stopped off at a local truss manufacturer's shop and bought a box of 2-in. by 4-in. truss plates. Nailed at the corners as shown above, the ⅜-in. prongs of each plate grabbed the neighboring sheets of plywood and held them fast throughout the job. And their low profile meant we never tripped over them.

—Matt Jackson, Rapid City, S. Dak.

DOUBLED JOISTS CAUSE LEAK

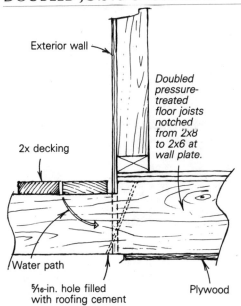

Exterior wall

Doubled
pressure-
treated
floor joists
notched
from 2x8
to 2x6 at
wall plate.

2x decking

Water path

⁵⁄₁₆-in. hole filled
with roofing cement

Plywood

My house is elevated on a pile foundation and has a deck
attached to one side. As shown above, pressure-treated doubled
floor joists extend beyond the exterior wall to carry the deck.
Whenever it rained, water would collect between the doubled
joists. Drawn by capillary action, the water would flow under the
house, where it would puddle atop the plywood nailed to the
bottoms of the floor joists.

To redirect the leaks, (I can't really stop them without ripping up
the deck and flashing the tops of the joists) I used a long ⁵⁄₁₆-in. bit
to drill holes in the seams between the doubled joists. I then used a
caulking gun to pump the holes full of roofing cement. The result is
a dam that keeps the water from getting to the plywood. Instead, it
drips beneath the deck itself.

—*Will Rainey, Galveston, Tex.*

RECYCLING A WALL

Garage end wall is lowered by crane onto its new foundation

I got a call from some people who wanted to expand their single-car garage to a double-wide. They didn't want to change the style of the garage—just stretch it. So I took some measurements, went home and started making a materials list.

We needed identical concrete blocks for the foundation, the same board-and-batten plywood siding, the same gable-end roof overhang and the same paint. The clients even wanted the shrub near the corner of the garage moved over 12 feet.

As I went through the parade of events in my mind—demolition, dump run, framing to-match-existing, siding to-match-existing, painting to-match-existing—it seemed crazy not to let the old end wall be the new end wall. After all, it was sheathed with plywood, which turned it into a giant gusset that would stay together nicely during a move. Around here, a crane costs about $100 an hour, but even with that expense, we figured we could save our client about $700. Here's what we did.

We built a new foundation, establishing its width and height by extending lines from the old one. We put the new anchor bolts in precisely the same place in the new foundation. To detach the end wall, we cut the nails at the corner studs and plates with a reciprocating saw, and pulled the siding nails where the sidewalls overlapped the corner studs. On the roof, we peeled back the shingles to a logical dividing line and cut the plywood roof sheathing. Before bringing on the crane, we added nails to the plywood siding wherever they were missing, and we left a few nails in place at the the corners to hold everything together.

As shown in the drawing above, the crane's cable was centered on the ridge of the end wall and secured to the top plate of the studwall on the interior side. With the cable pressure snug, we pulled the last few nails, and away she went. We slipped the wall over its new anchor bolts, plumbed and braced it, and filled in the empty spaces with new garage.

—Roger Gwinnup, Oxford, Iowa

THE SPREADER

Straighten spreader to push wall.

V-joint

16d both sides

2x4 or 2x6

Often during framing we need to align a stud wall to bring it plumb. Long interior walls or unsheathed exterior frames normally need slight adjustments, and trying to muscle them by hand doesn't always work very well. Faced with this situation, we use our spreader. It is simply two lengths of 2x4 or 2x6 assembled at midpoint by way of a V-joint, as shown in the drawing above. The two pieces are attached with a pair of 16d common nails, driven in line with one another allowing the pieces to rotate freely.

To find the correct length 2x for a spreader, measure from the top plate at the wall's end to the subfloor or ground at about a 30° angle. Standard 8-ft. walls need a spreader about 10 ft. long. Wedge one end of the spreader against the top plate. With a slight upward bend in the spreader, wedge its lower end against a block or a stake. Then simply apply slight downward pressure at the V-joint, and the top of the wall will move away from you with ease. Remember: the spreader is for making slight adjustments, not major ones.

—Stephen Major, Homer, N. Y.

POSITIONING STUD WALLS

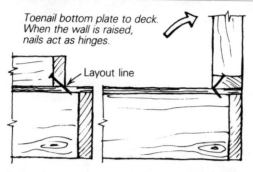

Toenail bottom plate to deck.
When the wall is raised,
nails act as hinges.

Layout line

Before I raise a stud wall, I find it helpful to toenail it with 16d
nails onto the layout line, as shown above. As the wall is hoisted
into position, the nails act as a hinge as they bend. The wall ends
up on the layout line with little need for adjustment, and no
protruding nails.

—Zack Mills, Olympia, Wash.

ANOTHER DECKING PERSUADER

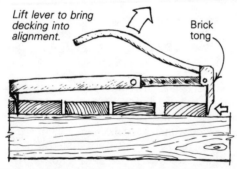

Lift lever to bring
decking into
alignment.

Brick
tong

Instead of using a pipe-clamp arrangement to align deck boards
(*FHB* #57, p. 26), I use a brick tong (drawing above). The stationary
end of the tong can be wedged between two nailed deck boards
and the pivoting end can be used to bring a board in, or, if the
anchoring end is moved back one board width, to push it out. This
takes less time than screwing and unscrewing the pipe clamp.

Because you need one hand to hold the tong handle, it helps to
start the nail before pulling on the board. You may also want to drill
a couple of new holes in the tong's adjustment bar to get the
spacing right.

—Phil Cyr, Dudley, Mass.

SPACING DECK BOARDS

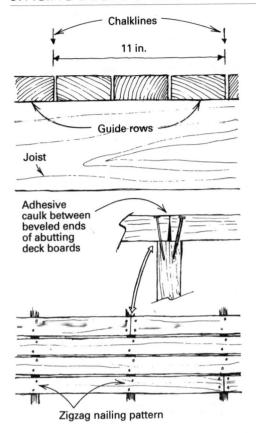

Chalklines

11 in.

Guide rows

Joist

Adhesive caulk between beveled ends of abutting deck boards

Zigzag nailing pattern

If you've ever laid decking using spacers to keep the rows parallel and straight, you know how frustrating things can become. Try as you may to prevent it, the rows have a way of wandering, and the bigger the deck, the more unsightly the wandering lines become.

For spacers to work, all the decking has to be exactly the same width. If not, the spacers will eventually move the material away from a straight line. And because even the best grade of decking isn't precisely the same width, I use a system of guide rows that avoids spacers altogether. Here's how it works.

I snap lines on the deck joists for every third row of decking, as shown in the drawing above. I determine the distance between the lines by adding the sum of three rows of decking plus three spaces. For example, I typically use air-dried Western red cedar 2x4s for deck boards and space them $\frac{3}{16}$ in. apart. Therefore, three rows of decking at $3\frac{1}{2}$ in. per row plus three gaps at $\frac{3}{16}$ in. equals

11$\frac{1}{16}$ in. I round this off to 11 in. If your decking is green, consider using smaller gaps to account for the greater shrinkage.

With lines snapped, I nail down every third row over the entire deck, taking great care to follow the chalkline exactly. Then I lay deck boards between the guide rows. By eye, and with the help of a thin steel bar or piece of wood, I align the boards until the gaps are equal. Because the spaces you're adjusting are confined to the joist right in front of you, your eye easily judges the distance between them, and the resulting job is surprisingly rapid and precise.

I nail each board in a zigzag pattern for two reasons. First, to me it looks better than a straight line, which is hard to keep perfectly straight anyway. Second, a straight nailing pattern is likely to split a joist, and that risk is eliminated with the zigzag pattern.

Where two pieces of decking butt on one joist, I undercut the decking ever so slightly—about $\frac{1}{16}$-in. total—and spread adhesive caulk over one end. The joint is tight to the eye and sealed against moisture soaking into the end grain. Then I nip the points off four 12d nails and angle them into the joist. This technique prevents splitting without having to drill pilot holes for the nails.

—David Bright, Lynden, Wash.

A CORNER-FRAMING METHOD

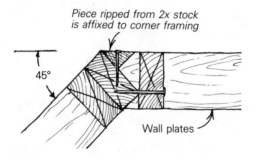

Piece ripped from 2x stock is affixed to corner framing

45°

Wall plates

Most methods for stick framing a corner that needs drywall backing work, but they require a bandsaw or a couple of passes on a 10-in. table saw in order to rip filler pieces made out of 4x4s. The illustration above shows how I use a 2x ripping made on a single pass on the table saw to accomplish the same thing.

—Brian Erdek, Califon, N. J.

ANOTHER CORNER-FRAMING METHOD

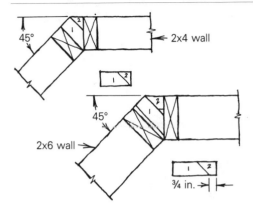

To make corner backing out of 2x stock, I simply set my circular saw on 45° and remove one edge of a 2x4 stud. Then I nail the ripping onto the stud, filling the corner void (drawing above). This method will also work with 2x6 studs if you make the rip ¾ in. from the corner instead of right on it. This method wastes no wood, requires no table saw and results in plenty of nailing surface.

—*Tom Mahony, Kona, Hawaii*

STAIR STORAGE

A lot of people use the space under basement stairs for storage. So do I. To eke out a little more storage space, I fastened cleats to the sides of the stringers beneath the stair treads. They carry shelves that give me a place for small items that would otherwise get lost in the big pile of stuff on the floor.

—*Robert M. Vaughan, Roanoke, Va.*

THE A-BRACE

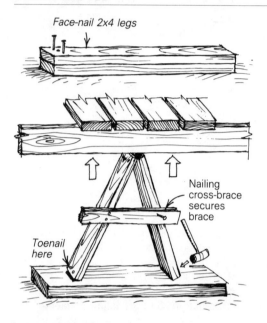

Face-nail 2x4 legs

Nailing cross-brace secures brace

Toenail here

I needed to lift a section of decking that had subsided a couple of inches, and I didn't have a jack handy to bear the considerable weight. Instead, I made the A-brace shown here out of a couple of short lengths of 2x4 and a 1x brace.

To make the A-brace, face-nail a pair of 2x4s at one end with three 16d nails through each side. The 2x4s should be 2 in. to 3 in. longer than the height you want to reach. Spread the unnailed ends of the 2x4s so that they form an inverted V beneath the appropriate beam and toenail one of the 2x4 legs to a secure 2x6 or 2x8 base. Forcing the legs together raises the load. Toenailing the other leg to the base, along with a cross-brace, secures it.

—Gerry Magid, Jamaica Plain, Mass.

SECURE CLOSET

Along with "more closet space," increasing concerns with security rank high with my remodeling clients. On several jobs I have combined the two needs into a "secure closet" for storing stereo systems and other valuables. If you are building a closet anyway, the extra effort and material is minimal. Existing closets can easily be beefed up.

Obviously, the door must be solid-core (preferably metal-clad). I hang it in an exterior jamb that has been blocked solid to the framing to eliminate any crevices that might accept the end of a crowbar. I prefer to install an outswing door to minimize the chances of someone kicking it in. This means the hinges have to have non-removable pins. I replace the regular hinge screws with 3-in. hardened screws (like those used for hanging cabinets) to make sure they reach well into the studs. I also use these screws instead of finish nails to hang the jamb. They go in fast with a screw gun. Then I install a good deadbolt with all its reinforcing hardware. Burglars have been known to kick in walls if they can't kick in doors. So as a final touch, I cover the interior walls of the closet with ¾-in. plywood.

Do all these precautions work? My local police department tells me that nothing will keep out a determined burglar if he is persistent enough—the best you can do is place as many barriers as possible in the way. When I moved to Oakland three years ago, the first thing I did was to build a closet of this type. Soon after, I had to leave town for a month. When I returned, my place had been broken into, but the secure closet had been frustrating enough that the burglars left empty-handed.

—Michael Mulcahy, Oakland, Calif.

SHEATHING CUTOUTS

Carpenters often cut out the plywood sheathing from door and window openings with a reciprocating saw, circular saw or even a chainsaw. I prefer to use a carbide-tipped pilot bit mounted in my router. This type of bit has a point on its tip, so it can be plunged into the center of the work. Above the tip is a pilot bearing, which will follow the framing as the cutters make a quick, clean cut in the sheathing.

—Michael Gornik, Nevada City, Calif.

HEADER IMPLANT

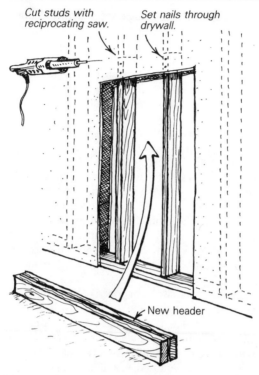

Cut studs with reciprocating saw.

Set nails through drywall.

New header

Not long ago my partner and I were working on a remodel that included putting in a new doorway through an existing wall. I was grumbling about what a pain it was because we'd have to cut out the extra drywall over the door and then turn around and patch it back in after installing the header. My partner gave me that superior look of "he-who-knows" and told me to watch.

First he cut out the drywall on both sides of the wall to the rough opening of the door. Then, with a ¼-in. punch, he set any drywall nails that were within 6 in. of the top of the opening, driving them through the drywall and into the studs. Next, he fired up his reciprocating saw and, piercing the drywall, cut off the studs 5½ in. above the rough opening (drawing above). After removing these studs, we slipped our header into the wall, followed by our trimmers.

We toenailed everything with 16d casing nails set through the drywall. To finish up, all we had to do was to fill a few holes and cuts, rather than having to mud and tape in a whole new chunk of drywall.

—*Tim Pelton, Fairfield, Iowa*

HEADER RETROFIT

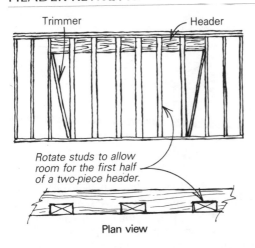

Trimmer — Header

Rotate studs to allow room for the first half of a two-piece header.

Plan view

Try this trick the next time you put a built-up header into an existing wall. First, use a reciprocating saw to sever the nails at the top and bottom of all the studs to be removed. Now rotate the freed studs 90° and align them on one side of the bottom and top plates, as shown above. Next, place trimmers (cut to their finished length) at an angle at either end of the opening. Raise one half of the header into place, and tap the trimmers into their vertical position (depending on the span, a mid-support for the first half of the header may be required). Now you can remove the original studs and add the second half of the header. Spike the two halves together, and you're done.

—M. Felix Marti, Monroe, Ore.

POLE-TOP TRIM

I once built a house with exposed poles, and their squared-off ends looked unfinished compared to the rest of the house. A large terracotta tile patio behind the house gave me an idea. Why not use the saucers that go with large clay flower pots as pole caps? I checked the nursery, and sure enough they had a big supply of very large saucers—up to 20 in. or so.

One of our crew members had been a lumberjack, so he put on his climbing gear and clambered up each pole. He attached each clay-pot lid to its pole top with asphalt mastic. The detail adds a nice finishing touch, and I think the saucers might slow down the end-grain checking process and resulting rot as well.

—Charles Miller, Kensington, Calif.

TRELLIS FLASHING

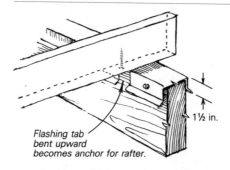

1½ in.

Flashing tab bent upward becomes anchor for rafter.

When I set about to repair an old trellis, I found that the rafters had developed dry rot around the toenails that affixed them to the support beam. To prevent this from happening again, I devised the combination flashing/anchor shown in the drawing above. First I had a metal shop make up a standard parapet-type flashing out of 18-ga. galvanized sheet metal. The legs and drip edge of the flashing are about 1½ in. wide. After laying out the positions of my rafters on the flashing, I used a sabersaw to cut 1-in. wide tabs in the flashing legs to correspond with the rafter layout. Bent upward, these tabs lie flat against the underside of the rafters. At each tab, I ran a 1-in. #10 sheet-metal screw into the bottom of the rafter. Now the trellis beam is protected by a flashing, and the rafters are anchored to it with fasteners that aren't exposed to the weather.

—Les Watts, Herndon, Va.

GOOD-NEIGHBOR POLICY

Our company does a lot of remodeling work in established neighborhoods, and to make sure that we get off on the right foot with the people living in close proximity to a project, we introduce ourselves with a preprinted card left in their mailboxes. At the top of the card is my name and contractor's license number, followed by our company's name, address and telephone number. Then the card goes on to say, "We are working on a remodeling project for your neighbor at: (space for the address). Please call us or stop by if we inadvertently create any problem or inconvenience for you. You are also welcome to drop by and visit the project during regular working hours."

The good will generated by this gesture helps to relieve tensions that a construction project can bring to a neighborhood. It has also led to a lot of other work for our company.

—Robert Malone, Berkeley, Calif.

SIMPLE STURDY GATE

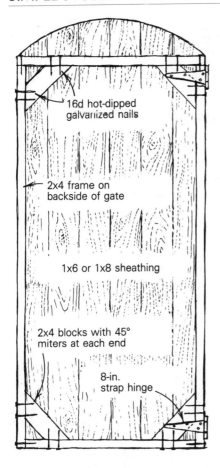

16d hot-dipped galvanized nails

2x4 frame on backside of gate

1x6 or 1x8 sheathing

2x4 blocks with 45° miters at each end

8-in. strap hinge

As a fencing and decking contractor, I'm asked to build a lot of gates. Instead of using the typical diagonal bracing, I place 2x4 blocks with mitered ends at each corner of the gate-frame (drawing above). There are two reasons for this: the 45° angles on the ends of the blocks ensure a square frame; and they provide good anchorage for the strap hinges that I use.

—Rick Fagnani, Alameda, Calif.

PICKET GAUGE STICK

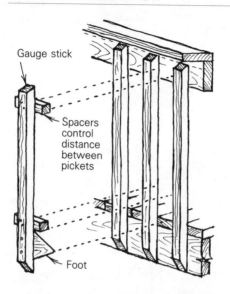

Gauge stick

Spacers
control
distance
between
pickets

Foot

The next time you're assembling railing pickets on a deck, try
using a gauge stick to speed assembly. As shown above, the gauge
stick fits between pickets. Spacers at its top and bottom make it
easy to align the next picket, while the gauge stick's foot rests on
the decking.

—Louis J. Fritz, Medford Lakes, N. J.

2
FOUNDATIONS, MASONRY AND TILE

CRAWL-SPACE CONCRETE

We had a job that included the excavation of some fair-sized pier footings under an existing house, and when it came time to pour we were looking at the unpleasant task of moving a half-yard of concrete in buckets while crawling on our bellies. We'd done this before, and were resigned to the inevitability of it all. But this time I saw a way to simplify the operation by using a 4-in. by 14-in. heater register installed in the floor about 3 ft. from our pier excavation.

From below, my colleague untaped the galvanized boot that led to the register, and put a bucket under the opening. Up top, I taped a protective layer of cardboard to the floor near the register, and made a cardboard funnel to direct the concrete through the opening as I shoveled it in. We worked the job with a pair of buckets. I filled one while Gary emptied the other one. We finished in a jiffy, wiped the boot with a rag and taped the duct back in place.

—*John Campbell, Santa Cruz, Calif.*

CONCRETE FUNNEL

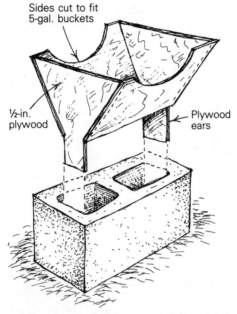

Sides cut to fit
5-gal. buckets

½-in.
plywood

Plywood
ears

Recently we had to fill with concrete the cells of some concrete-block piers. We naturally wanted the job to go quickly and neatly—no slopping wet concrete on the sides or at the base of the blocks. To help us with the pour, we used the plywood funnel shown above. It was held steady atop the blocks by the plywood ears extending into the block cells, and the curved cutouts allowed us to prop our buckets on the funnel edges without worrying about missing the mark.

—Gregory D. Lang, Cedar Key, Fla.

CHALK LINES ON WET CONCRETE

What do you do when you're ready to start snapping chalk lines for the walls on a new slab, but the slab is wet and more rain is on the way? You could answer, "Head for the coffee shop." But that's not it.

I go to the local rental yard and get a "weed burner." It's simply a 5-gal. propane tank with a 6-ft. hose and a torch nozzle. Next, I use the burner to dry a 6-in. wide path on the slab in the general area of a wall. Then I snap the wall lines and spray them with quick-drying lacquer (it comes in an aerosol can). After a couple of minutes the line is there to stay, regardless of rain or constant sweeping.

—Gil Meador, Jacksonville, Ore.

GIANT-CONTOUR GAUGE

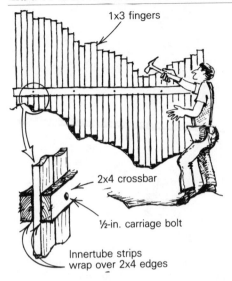

1x3 fingers

2x4 crossbar

½-in. carriage bolt

Innertube strips
wrap over 2x4 edges

The toughest part about pouring a concrete wall atop a rocky
ledge is getting the forms to match the undulating surface of the
rocks. A bad fit under these circumstances can mean a lot of lost
concrete and some heavy messes to clean up. Faced with this
challenge, we took our inspiration from a tool usually put to use by
trim carpenters—the contour gauge (drawing above).

Like the smaller version, ours has individual fingers held in place
by a crossbar. The fingers can be moved independently to mimic
an uneven surface. But instead of tiny stainless-steel pins, the
fingers in our version are 1x3s. They are held in place by a pair of
2x4s, which are lined with innertube to keep the fingers from easily
slipping. Five ½-in. carriage bolts hold the thing together.

To use the gauge, we adjusted the 2x4 crossbar so that it was
level, and then tapped the fingers down until they touched the
ledge. When all the fingers were touching the ground, we tightened
the bolts to make sure they wouldn't move. Then it was a simple
matter to transfer the profile of the ground to our form plywood
and cut out a perfect form—first time.

—*Dan Tishman, Andy Williamson, Damariscotta, Maine*

TRANSIT MIXER

Steel pin
secures
handle to
pipe sleeve

Go-cart wheels

I got tired of muscling my cement mixer around the job site, so I mounted a pair of go-cart wheels on one end (shown above). To lift the mixer for transport, I used a pair of handles made from ½-in. galvanized pipe. They fit into four sleeves made of ¾-in. black pipe that were brazed to the mixer's frame. The handles are removable, but to make sure they stay put, I connected each handle to a sleeve with a steel pin.

—*Joseph A. Fletchko, Ocean City, Md.*

FINE TETHERED FRIEND

Running a vibratory compactor is a boring job, but it must be done before pouring a concrete slab. The last time I was faced with a day behind the bouncing handlebars, I decided to automate the job. Vibratory compactors are designed to creep slowly forward in a straight line. My job was to get it to turn without me. As shown in the drawing above, I tied it to a stake driven in the center of the excavation. Each revolution of the compactor shortened the leash, resulting in circles of ever-decreasing diameter. This left me free to take care of other tasks on the site as I kept track of the machine out of the corner of my eye. I still had to finish up the corners by hand, but the trick saved me a lot of time and wear and tear on the wrists and elbows.

—Mark White, Kodiak, Alaska

STANDOUT STAKES

Concrete patios, driveways and paths have to be sloped to promote good drainage, and I've poured more sloped concrete than I care to recall. The challenge with this kind of work is getting the levels right, and that means installing plenty of elevation stakes. But elevation stakes can be tough to drive into rocky soil or dry-clay soil. The stake ends up split or burred. And to make matters worse, wood stakes are easily lost in the excitement that often accompanies a concrete pour.

On my last sloped slab, I used stakes made of ½-in. copper water pipe. Not only were they easy to drive into the hard ground, but they were also equally easy to see as I raked the concrete and screeded it to its final elevation. Each pipe stood out as a perfect, dime-sized black spot against the light-gray concrete. Evidently the ½-in. dia. hole is too small for the average piece of aggregate to clog, and the concrete cream doesn't have the viscosity necessary to span the hole. As soon as the concrete has been screeded to its final elevation, you can either pull out the stakes or drive them with a length of pipe beyond the bottom of the slab.

—Tony Toccalino, Milton, Ont.

STAKE PULL

On a recent foundation job, I was stunned to discover that we didn't have any grease to coat our steel stakes with. There wasn't time to send out for some, so we looked around the site for some other material that would keep the stakes from becoming part of the foundation. Our roll of filter fabric came to the rescue. We protected each stake with a couple of wraps of the fabric, and held the stuff in place with tie wire. Pour complete, the stakes came out with little effort.

—Matthew Bronson, Oakland, Calif.

REBAR SHEATHS

Concrete footings frequently require vertical rebar stubs projecting upward to anchor walls that will be built at a later date. Because they've been cut, these stubs usually have very sharp ends. As soon as vertical bars are in place, I cover their tips with a small piece of duct tape, rendering them relatively harmless during the pour. The tape also protects the mason's hands when it comes time to set the wall blocks.

—Bob Jewell, Kalaheao, Kauai, Hawaii

HOMEMADE REBAR BENDER

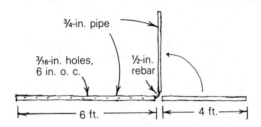

Recently I was forming footings for an addition and needed to rent a bender to put angles in the ½-in. rebar. It wasn't available at the local yard, so I improvised the simple tool shown above. It works well enough that for me, renting rebar benders is a thing of the past.

My bender consists of two pieces of ¾-in. steel pipe—one 6 ft. long, the other 4 ft. long. In use, the 6-ft. length stays on the ground; the 4-ft. length is the lever. If I'm bending special pieces, I measure the rebar from end to bend, mark it, and slide the pipe over the rebar until the mark is between the two pipes. Lifting on the lever to the desired angle makes the bend.

For repetitive bends, I drilled a series of ³⁄₁₆-in. diameter holes, 6 in. apart, in the 6-ft. length of pipe. An 8d duplex nail in one of the holes serves as a stop.

—Dan Jensen, Tigard, Ore.

RETRO-FLASHING A CONCRETE STOOP

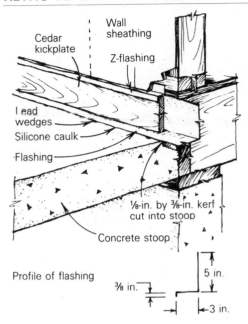

Cedar kickplate

Wall sheathing

Z-flashing

Lead wedges

Silicone caulk

Flashing

⅛-in. by ⅜-in. kerf cut into stoop

Concrete stoop

Profile of flashing

⅜ in.

5 in.

3 in.

The entry porch to my client's house is on the west side of the building, where the rain is sometimes driven hard by the wind. The porch floor is a reinforced-concrete slab that doubles as the roof of a small fruit cellar. Whenever the rain came from the west, water would find its way into the cellar by sneaking past the cedar kickplate under the doors and down the rim joist. To remedy the problem, I began by prying the kickplate away from the wall and out from under its Z-flashing. Then I used a masonry blade mounted in my circular saw to cut a ⅛-in. wide by ⅜-in. deep kerf in the concrete stoop. This took several passes. Next I bent a piece of dark brown aluminum fascia material into the flashing profile shown in the drawing above. Before I let the ⅜-in. lip of the flashing into the stoop, I swept the kerf clean and filled it with a bead of silicone caulk. Then I inserted the flashing, and secured it in the kerf by driving small lead wedges (cut from salvaged lead pipe) every foot or so along the lip of the flashing. I finished by applying another bead of silicone to the kerf and by tucking the cedar kickplate back under its Z-flashing.

—Tim Herrling, Auburn, N. Y.

ANCHOR-BOLT SPACERS

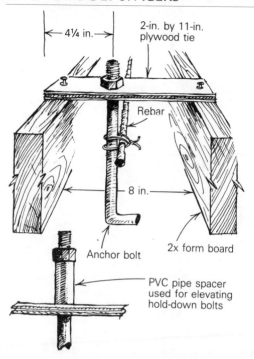

4¼ in.

2-in. by 11-in.
plywood tie

Rebar

8 in.

Anchor bolt

2x form board

PVC pipe spacer
used for elevating
hold-down bolts

Whenever I build forms for stemwall foundations, I like to have all my anchor bolts in place before the pour. This ensures proper alignment in the center of the mudsill, correct elevation (no chiseling the mudsill to accommodate the washer and nut) and accurate spacing so as not to interfere with my joist layout. In addition, this method guarantees a good bond between the concrete and the bolt, and eliminates the problem of having to poke anchor bolts into concrete that has already begun to set up.

To position the bolts, I use 2-in. by 11-in. rippings of ⅝-in. or ¾-in. plywood. As shown above, the rippings also act as ties between opposing 2x form boards, and they can be reused many times. I drill holes in the ties to accept ½-in. anchor bolts. The holes are centered 4¼ in. from the outside end of the ties, placing a bolt squarely in the middle of a 2x6 mudsill on an 8-in. stemwall. The bolts can be inserted through the holes with the nuts already on.

I typically tie each anchor bolt to the rebar to make sure the bolt stays put during a pour. Another way to keep the bolt in place as the concrete is poured around it is to slip a length of ¾-in. pipe over the bolt threads. This also keeps the concrete off the threads.

Here in California we have to install beefy hold-downs connected to large-diameter foundation bolts to keep buildings together during earthquakes. The bolts often have to be held exact distances above the finished stemwall. I use the plywood strips in conjunction with PVC pipe spacers (detail drawing, facimg page) to make sure the top of the bolt ends up just where I want it.

—Yon Mathiesen, Soquel, Calif.

BEVELED RISER FORMS

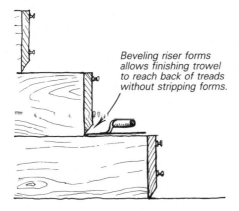

Beveling riser forms allows finishing trowel to reach back of treads without stripping forms.

When building forms for concrete steps, use 2x material to form the risers as shown in the drawing above, and cut a bevel along the bottom edge. This allows room for the finishing tools, such as the float and trowel, to reach to the back of the tread without having to wait for the forms to be stripped. The result is a tread with a uniform finish from the nosing to the riser.

—Walter H. Chandler, Richmond, Va.

BASEMENT-WALL ANCHORS

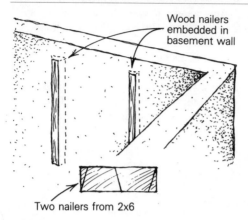

Wood nailers embedded in basement wall

Two nailers from 2x6

Sooner or later a basement wall ends up with a row of shelves on it or gets buried behind layers of insulation and drywall. The nailing strips shown above will work to anchor these things, and can be easily included in the basement walls as they are poured. I make a pair of nailers out of a single 2x6, ripped with beveled edges as shown in the detail. During the pour, I make sure to vibrate the concrete alongside the nailers to eliminate voids around their vertical edges.

Also, I don't bother with pressure-treated lumber for these nailers. If they stay dry, the nailers will last indefinitely. And a properly constructed basement wall is a dry wall.

—Burleigh F. Wyman, Whitefield, N. H.

STEP-DRILLING IN MASONRY

When I build decks or additions, I often have to fasten a ledger with lag bolts and lead shields to an existing masonry wall. This process can be maddening because a few of the holes always end up a little off layout. The problem is that a big masonry drill bit has a tendency to wander off its mark.

To avoid this problem, I first drill my holes in the ledger, then mark their center points on the masonry. Then I start drilling the holes in the masonry with a ⅛-in. bit, followed by a 5/16-in. bit and finally a ½-in. or ⅝-in. bit, depending on the diameter of the lead shield that I'm inserting. This extra effort takes only a few minutes, and it consistently yields accurate results.

—John A. Neer, Alexandria, Va.

MOLDINGS ON MASONRY WALLS

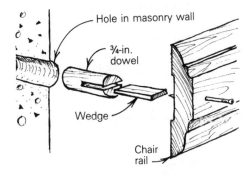

The drawing here illustrates the method that I use to apply trim pieces, such as chair railings, to a concrete, brick or block wall. I do it this way because it allows me to use finish nails to secure the work to the wall. Then the nail holes can easily be filled to match a natural or stained finish.

First I drive nails into the railing so that their tips just begin to emerge on the backside. Then I hold the railing in place on the wall and drive the nails far enough to make marks. Next I drill ¾-in. holes in the masonry, using the nail marks as centerpoints. I fill each hole with a piece of ¾ in. dowel that has been kerfed on one end with a bandsaw. If I'm working on a concrete block wall with open cells, I make sure that the dowel is long enough to bear against the far inside wall of the block.

Once the dowels are wedged in place and trimmed flush with the wall, I affix the railing to the wall by driving the nails into the dowels.

—Jim Stuart, Covina, Calif.

NONROLLING STONES

River rocks make beautiful veneers on walls or fireplaces, but their smooth surfaces make it difficult to anchor them firmly to a substrate, especially if the mortar joints are deeply raked. That's why I "prime" the backside of each rock to improve adhesion.

First I prepare a very thin (milklike) slurry of water and plastic cement. Plastic cement is sold at any well-stocked building-material supply house that carries masonry products. It's not as strong as portland cement, but it has a plasticizer added to it that makes it stickier. Plastic cement is also a little more expensive than common cement.

After soaking the stones in plain water, I place them face up in a shallow pan containing the slurry mixture—about 3 in. deep or so. I want only the backside of the stones to touch the slurry. After a five-minute soak, the stones are ready to be set with ordinary mortar. Using this method, I've never had a stone come loose.

—Bill Hart, Templeton, Calif.

TRENCH PLUGS

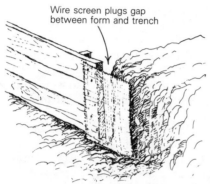

Wire screen plugs gap between form and trench

I recently poured grade beams and stemwalls for a house foundation on a hillside lot with stony soil. When the wooden form boards butted up to a trench or embankment, the irregular surface of the ground made for ragged gaps that were difficult to seal against the pressures of several feet of wet concrete.

To solve this problem I used roofing nails to tack ¼-in. by ¼-in. wire screen over the ends of the form boards and across each gap (drawing above). The weight of the concrete forced the screen to sit tightly against the bumpy ground, and the screen kept the concrete inside the forms. In those places where the concrete pressure was greater or the gaps wider, I simply doubled up the wire.

—Glen Carlson, San Diego, Calif.

ROLLING ROCKS

Whenever I have to move heavy rocks as I go about my business as a stonemason, I reach for my rollers. I used to put pipes or dowels under the rocks to roll them from place to place, but I found that even a small pebble would put the skids on progress. My local wrecking yard held the solution—coil springs from old pickup trucks. They will roll easily across a stony surface, pushing aside any pebbles that get in the way. It's good to have four coils for stone moving. Put the last one back in front when the stone rolls beyond it.

–*Stephen Kennedy, Biglerville, Pa.*

FOUNDATION-WALL REINFORCEMENT

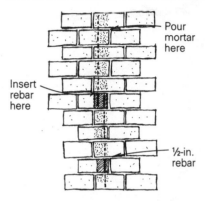

Pour mortar here

Insert rebar here

½-in. rebar

As the old oak tree next to my foundation continued to grow, it exerted enough force on the hollow concrete blocks to create horizontal cracks along the mortar lines. I had the tree removed, but before repairing the cracks I wanted to restore the wall's structural integrity—without dismantling it.

I began my work by removing an 8-in. by 5-in. section of block, about halfway up the wall (drawing above). That gave me access to a cavity that ran through the cells of interlocking blocks from the footing to the top of the wall. Next I inserted a pair of slightly bowed pieces of ½-in. rebar—one up and one down. I wired them together where they overlapped at the hole.

Near the top of the wall, I excavated another hole for pouring mortar into the cavity. I inserted a hose with a spray nozzle adjusted to a fine mist to moisten the interiors of the blocks before filling them with mortar mix. A short length of 4-in. PVC pipe and a cut-up milk jug made a good funnel to direct the mortar into the wall. Using the rebars as agitators, I was able to fill the cavities thoroughly with mortar.

I poured these columns in every other cavity along the worst section of the wall, and every fourth cavity in the portions of the wall where the problem wasn't as serious. Covered with a new coat of plaster, the wall now has its original, crack-free finish.

–*Donald L. Anderson, Takoma Park, Md.*

POURED-IN-PLACE CHIMNEY CAP

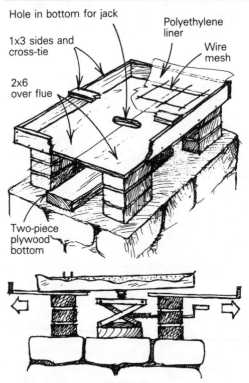

Hole in bottom for jack

Polyethylene liner

1x3 sides and cross-tie

Wire mesh

2x6 over flue

Two-piece plywood bottom

To remove formwork, pull 1x3 ends
and lift cap with jack.

We needed a chimney cap to keep the rain from pouring down our
fireplace flue, but the estimate we got to install a 2½-ft. by 3½-ft.
2-in. thick piece of stone atop our chimney was outrageously high.
I decided to pour a concrete cap for the chimney, knowing that the
primary challenge lay in getting the thing to the top of our 2½-story
house. My solution was to pour the cap in place, and then remove
the formwork with the help of my car jack.

As shown above, I made a rectangular form with a two-piece,
½-in. plywood bottom and 1x3 pine sides. A cross-tie held the
longer opposing sides together. Near the center of the bottom I cut
two U-shaped notches. They would eventually allow me to raise
the cured cap enough to remove the formwork.

I supported the form with a stack of mortared bricks under each
corner. Then I placed a 2x6 over the flue, and jockeyed the car jack
around on it until it was under the holes in the bottom of the form.
To make it easier to remove the formwork, I lined the form with

polyethylene. The poly also kept the concrete from escaping through the hole in the bottom of the form.

When the form was ready, I mixed the concrete at ground level and hauled it up a series of ladders to the form. I reinforced the concrete with wire mesh, and placed anchor bolts in the wet concrete to secure the base of a lightning rod. I let the cap cure a few days under wet burlap bags, then used the jack to lift the cap enough to slip out the plywood. At the same time, I slipped a ping-pong-ball sized blob of epoxy under each corner to bond the cap to the bricks. Eleven years have gone by, the cap hasn't moved and the hearth remains dry.

—John D. Hallahan, M.D., Media, Pa.

BLOCKS FROM BOX FORMS

Joint compound box as pier block form

Taper top for drainage

When my stepfather was setting the piers for the deck around his new house, he discovered that he needed more pier blocks. Glancing around, he spied a pile of cardboard joint-compound boxes left by the drywall crew. The boxes were about the same size as a precast pier, so he folded back the box flaps top and bottom and filled one with concrete to test its ability to retain the concrete while it set. No problem. For pads, he used squares of redwood 2x6s with 16d galvanized nails driven into their bottoms. As shown in the drawing above, the concrete should be tapered away from the pads for drainage.

—Ken Morrill, Corralitos, Calif.

3

FLOORS AND STAIRS

PATCHING A HARDWOOD STRIP FLOOR

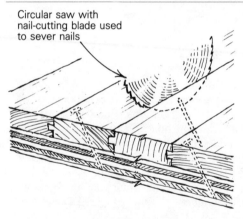

Circular saw with
nail-cutting blade used
to sever nails

To make a convincing patch in a hardwood strip floor, you've got
to stagger the ends of the adjacent strips. Otherwise, the patch will
look like a poorly concealed trap door. The problem is, how do
you accurately and efficiently make numerous cuts in hardwood
strips while they are already in place? The drawing above shows
how I recently solved this dilemma. The tools required are a
plunge router with a ¼-in. or ⅜-in. straight bit, a circular saw with a
nail-cutting carbide blade, safety glasses and a face shield, a chisel
and a nail-finder (you can use either the swinging-magnet kind, or
one of the coil types powered by a battery).

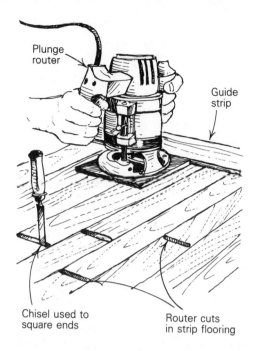

Plunge
router

Guide
strip

Chisel used to
square ends

Router cuts
in strip flooring

First, find out which edge of the strips have the tongues, and therefore the nails. If you're working from a visible end, this will be obvious. Otherwise, find a spot in the middle of the patch that doesn't have any nearby nails, make a couple of cuts across a strip and pry it out. Next, run the nail finder down the edge of each strip and mark the locations of the nails. Now you can lay out a series of staggered cut lines for the new strips that avoid any nails. To sever the nails, make plunge cuts on the waste side of the strip flooring—be sure to wear the safety goggles and face shield during this operation.

I used my router, guided by a strip of wood, to make perpendicular cuts across the strips (shown above). The width of the cut gave me enough room to get the end of a pry bar under the strip that needs to be pulled up. At this point, a chisel is sometimes handy for splitting the tongue off a strip that is reluctant to come out. I also use the chisel to square the ends of the slots. Chamfering the leading bottom corners and edge (plus a little patience) will help as you insert the replacement strips.

—*Ted Garner, Chicago, Ill.*

MUDBUCKET HELPERS

Recently I came up shorthanded on a job site where we had planned to lay a ¾-in. T&G plywood deck. Knowing this can be a frustrating job even with help, I was loathe to start it by myself. But there was no other work to do. Spying a trio of full joint-compound buckets, I decided to go ahead and put down some decking single-handedly. After nailing down the first row, I used the buckets as weights to hold down the tongues of the second row as I drove sheets home with a 2x4 and a sledge. This method worked so well that I got the entire deck laid before my tardy helpers arrived.

—Dennis Darrah, Montpelier, Vt.

PLANNING FOR TILE

As a kitchen and bathroom remodeler, I install quite a bit of ceramic tile. I've found that if my project is more than a basic square or rectangle, a scaled sketch of the countertop, floor or wall to be tiled can save a lot of time and trouble when the job rolls around.

With my project dimensions in hand, I use the ¾-in. scale on my architect's rule to draw the outline of the area to be tiled on a sheet of tracing paper. On another sheet of paper and with the same scale, I draw a grid to match the size of the tile selected for the job (in fact, I have several grid sheets to match common tile sizes— 12x12, 8x8, 6x8 and 4x4).

Grid drawn, I can get a quick and accurate picture of the job by overlaying my drawing of the project on the grid. The tiles show through the tracing paper, and I can move my drawing around until I get the best possible layout. With this approach, I can usually spot and solve any layout problems before I even start working. It's also a piece of cake to figure my materials by simply counting up the number of tiles the layout requires.

—Herrick Kimball, Moravia, N. Y.

SILICONE APPLICATOR

We coated the grout on our tiled hearth with silicone to protect it from staining and to make cleaning easier. Normally, this solution is applied with a sponge brush. However, I've found a better way. I put the solution in a glue bottle whose application tip has a very small hole. I then flood the grout line using this bottle. The grout soaks up the solution very quickly, so I keep applying the solution until it no longer soaks into the grout. This technique makes for a much quicker and more thorough job. If any excess silicone remains on top, I wipe it off with a clean cloth.

—*Neil Hartzler, Colorado Springs, Colo.*

MARKING ON TILE

As a professional tilesetter, I've occasionally used a grease pencil to mark my cuts. But I prefer another marker. My favorite is an aluminum knitting needle, which will leave a legible mark on all but the glossiest porcelains. And unlike the grease pencil, it leaves a very precise line. When the needle dulls, I sharpen it with a file, a sander or a grinder. I've been using the same needle for over 20 years, and I haven't begun to wear it out.

—*Ken Morrill, Corralitos, Calif.*

TILING NEAR TRIM

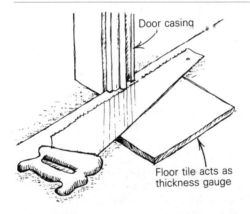

Door casing

Floor tile acts as thickness gauge

You can use a contour gauge to mark the irregular cuts required to notch a tile around a door casing, but there is a much simpler method. Place a tile next to the casing and use it as a thickness gauge for a handsaw as you trim the bottom of the casing (drawing above). Casing trimmed, the tile will fit under it without any complicated cuts. This trick also works with resilient tiles, but when using one, be sure to turn it upside down to keep from marring its finished surface.

—*John Molnar, Moorestown, N. J.*

CUTTING HOLES IN TILE

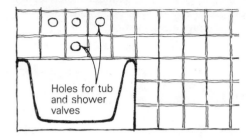

Holes for tub and shower valves

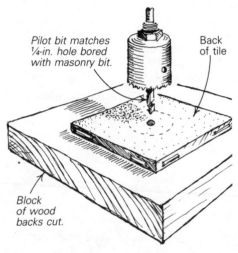

Pilot bit matches ¼-in. hole bored with masonry bit.

Back of tile

Block of wood backs cut.

While tiling a bathroom I ran into problems when I had to cut holes in new tiles for the tub and shower valves. I couldn't use my nippers to nibble away the tiles because each hole landed near the center of a tile. Instead, I used a 1⅜-in. dia. hole saw—the kind normally used to cut holes in wood (drawing above).

After laying out the centers of the holes on the backs of the tiles, I used my ¼-in. carbide-tipped masonry bit to drill pilot holes in each tile. Then I put the tiles face down on a block of wood, and used the hole saw to finish the job. The wood block reduced chipping on the glazed face of the tile and gave the pilot bit something to bite into. I must admit, this operation dulled the hole saw beyond resharpening. But at less than $5, it was worth it.

—Paul M. Dandini, Boston, Mass.

CLAMPING A STAIRCASE

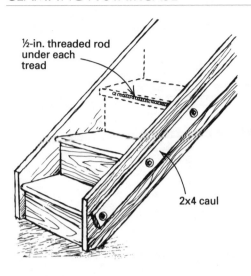

½-in. threaded rod under each tread

2x4 caul

Last winter I was involved in a complete rebuilding of a large house in Hyde Park, Chicago. The main staircase had been moved during an earlier remodeling. In that move, the wall-side stringer had been affixed to the studs with nails, which drew the stringer away from the housed treads. After about 50 years of use, along with the typical settling and seasonal movement that a house undergoes, some of the treads were barely supported in their grooves.

Once we began to rebuild the house, we had to pull the staircase back together. I began by drilling ⅝-in. holes below each tread. On the exposed, room-side stringer, I clamped a 2x4 the full length of the stringer to prevent tearout when my bit came through and to act as a full-length caul. I drilled corresponding holes in the wall-side stringer and removed the plaster under it to allow access to the backside of the stringer.

After removing the old wedges from the treads and the risers, I ran ½-in. threaded rods through each hole, as shown in the drawing below. Then I spun some washers and nuts onto the ends of the rods and slowly drew them down, clamping the whole stair together again. I finished the job by adding new wedges, epoxied in place, and filling the holes on the exposed side of the stringer with plugs that matched the original oak.

—Felix Marti, Ridgway, Colo.

FITTING STAIR TREADS

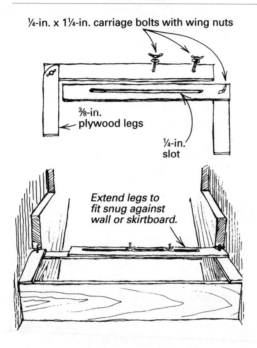

¼-in. x 1¼-in. carriage bolts with wing nuts

⅜-in. plywood legs

¼-in. slot

Extend legs to fit snug against wall or skirtboard.

Cobbled together out of scrap plywood, the jig shown above makes it easy to fit stair treads between a pair of walls or skirtboards. To use the jig, loosen the wing nuts and lay it across the tread cuts of the stair stringers, with its back tight against the riser cuts. Then extend the legs so that they're snug against the walls or skirtboards and tighten all the nuts. Remove the jig carefully, place it onto the tread stock and scribe the end cuts on the stock. When cutting, leave just a trace of the cutline for a perfect fit.

—Robert Plourde, Atlanta, Ga.

FOAM SHIMS

We recently devised a method for trimming enclosed staircases that resulted in gap-free intersections between the treads, the risers and the skirtboards. Here's how we did it. Before installing the ¼-in. veneered plywood skirtboards, we stapled a layer of ⅛-in. foam to the walls behind the skirts. The foam allowed enough give so that we could then install all the treads and the risers—cut ¹⁄₁₆ in. oversized—and get a perfect fit all the way up the staircase. If a small gap turned up somewhere, we drove a thin wedge between the plywood and the foam. Then we capped the exposed edge of the plywood with a piece of solid wood trim.

—Howie Roman, Rick Bockner, Whaletown, B. C.

LAG-SHIELD LAYOUT

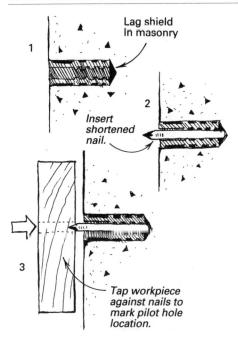

I had to replace some wood treads and risers on concrete steps, and I wanted to use the existing anchor holes. Rather than try to transfer the positions of the anchors to the new boards by measuring, I used the method shown in the drawing above. First I tapped the new anchors into their holes. Then I selected a nail that fit snugly into an anchor and cut it to a length that protruded just past the plane of the concrete. I put a shortened nail into each anchor, lined up the new pieces and tapped on them over each anchor. The resulting indentations perfectly registered the holes needed in the new treads.

—Al Lemke, Hopewell Junction, N. Y.

ALIGNING HOLES IN TUBING

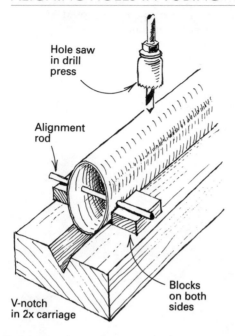

Hole saw in drill press

Alignment rod

V-notch in 2x carriage

Blocks on both sides

I made the balustrade for my new house out of wood and metal. The handrail is aluminum tubing while the turned balusters are red oak. The tapered ends of each baluster fit into holes in the underside of the tubing, and therein lies the challenge. How do I make sure the holes in the tubing all end up in a straight line?

I used a ½-in. hole saw mounted in my drill press to cut the holes. To cradle the tubing, I cut a V-notch in a length of 2x to serve as a carriage. Then I tilted the table of my drill press to the required 38° angle and clamped the carriage to it. Near the end of the tubing, I drilled a couple of ¼-in. holes for an alignment rod. As shown above, the rod bears on a couple of blocks. Nestled in this manner, the tubing can't rotate, and my holes end up straight and true.

—*Neil W. Momb, Issaquah, Wash.*

JOINING STAIR RAILS

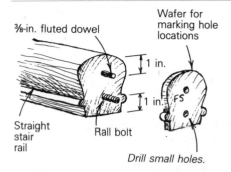

⅜-in. fluted dowel

Wafer for marking hole locations

1 in.

1 in.

FS

Straight stair rail

Rail bolt

Drill small holes.

In the past I joined sections of straight stair rail to fittings in the same manner described by Sebastian Eggert in his article on installing manufactured stair parts in *FHB* #39 (pp. 44-48). I used the rail bolt that came with the fitting, with a hex nut instead of the furnished star nut. But frequently during installation of the assembled rail sections, the glue line would break from the shock of coaxing the rail onto the balusters with a rubber mallet, or from the twisting action of aligning the rail with the newel posts. Then I'd have to reglue and contour the joint all over again.

I remedied this situation by adding a ⅜-in. dowel to the joint, as shown here. I use a wafer, cut from a section of rail, to mark the location of both the hole for the rail bolt and the dowel hole. To make sure I don't compound any error in the placement of the holes, I mark an "FS" on one side of the wafer to indicate the "fitting side," and I take care to orient it correctly. In addition to strengthening the joint to withstand the strain of installation, the dowel resists the torquing action of tightening the nut during assembly, helping to maintain proper alignment of the parts. Since I started using this method, I haven't had to rework a single joint.

—D. B. Lovingood, Suffolk, Va.

STAIR-RAIL SANDING BLOCKS

The other day I was working on a stair railing-to-turnout connection (they never match), and I needed a sanding block. I looked around for a suitable block, and it was actually at my feet. There I found a 4-in. piece of scrap handrail, just waiting to have a piece of sandpaper wrapped around it. It fit perfectly in my hand. When I started fitting the handrail pieces together at the point where the rail curved upward at a landing, I used a scrap piece of the gooseneck for a block. It had just the right radius.

—Richard Sena, Jamestown, N. Y.

PIVOT PLATES ON CARPET

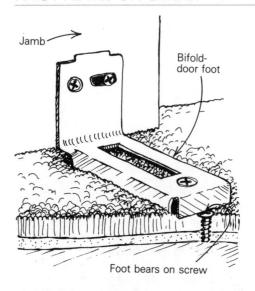

Jamb

Bifold-
door foot

Foot bears on screw

It can be tough to install a bifold-door pivot foot over a carpeted floor—especially when the carpet is plush and the pad is thick. Even though the foot is affixed to the jamb, the horizontal leg doesn't have the solid bearing necessary to support the weight of a door. So people usually put a little block of wood under the foot, hoping that it will stay put and not be too noticeable.

The drawing above shows the method I've developed to eliminate the little block of wood. First I secure the foot plate to the jamb at the correct height with a screw driven partway into the oblong hole in the vertical leg. Then I swing the foot out of the way, and drive another screw through the carpet and into the subfloor where the end of the foot will bear on the screw's head. I adjust this screw until its depth is right, then I run the screws into the side jamb on the vertical leg. The foot now has solid bearing, and the depth screw is practically invisible.

—*Glenn J. Goldey, Colwyn, Pa.*

4

WALLS
AND CEILINGS

RETURN BACKING

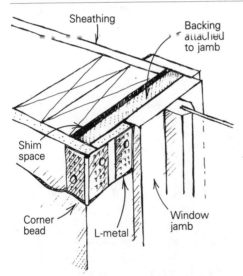

Sheathing

Backing
attached
to jamb

Shim
space

Corner
bead

L-metal

Window
jamb

My partners and I use drywall returns similar to the ones that
Dennis Darrah wrote about in his article *(FHB* #63, pp. 50-53). We
found, however, that some of the savings that should accompany
this budget-conscious detail were negated by the time spent
affixing furring strips to the rough framing so the drywall would
reveal evenly against the windows. To save time and get

consistently good results, we developed the detail shown in the drawing on p. 61.

First we rip inexpensive 1x pine to the depth of the framing members plus the exterior sheathing. Then we attach these backing strips to the rough side of the window jambs. To accommodate the increased dimensions of the windows, we make our rough openings 1½ in. wider and ¾ in. taller. After installing the windows, we shim the pine against the framing so that it is firm enough to receive drywall and corner bead.

Before we came up with this system, prepping window openings for drywall was a dreaded task, and the reveals were never as good as we wanted them to be. Now it's a piece of cake and the results are beautiful.

—Billy Guild, Hawley, Mass.

SOLO DRYWALL HANGING

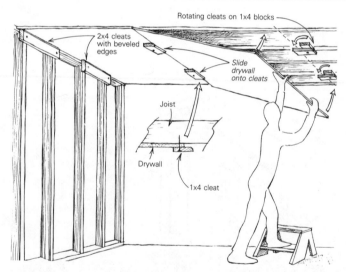

I've had to hang a fair amount of drywall on ceilings by myself, but rather than prop the panels in place, I prefer to use cleats.

As shown above, one edge of a drywall panel is supported first by a pair of cleats with beveled edges. When I'm working next to a wall, I use 2x4 cleats tacked to the studs with duplex nails. In the field, I screw 1x4 cleats through the prior sheet of drywall and into the joists.

The opposite edge of the drywall is supported by a pair of rotating cleats made of 8-in. long pieces of 1x4. Each cleat is affixed with a single screw to a 1x4 block screwed to the joists. To make sure I've got enough room to maneuver the drywall, I put the blocks 48½ in. from the edge of the prior panel. Once I've got the edge of the drywall on the fixed cleats, it's a simple matter to raise it the rest of the way and spin the two rotating cleats to hold the sheet aloft for final positioning. My system goes a long way towards saving my back.

—Fred Grosser, Redwood City, Calif.

PIPE-CLAMP DRYWALL LIFT

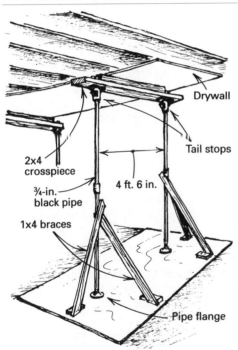

Drywall

Tail stops

2x4 crosspiece

¾-in. black pipe

4 ft. 6 in.

1x4 braces

Pipe flange

When it came time to raise high the drywall for my ceiling, I put together a pair of quickie drywall lifts out of some pipe-clamp fixtures and 4-ft. lengths of ¾-in. pipe that I had on site. First I removed the stationary heads from four clamps, leaving only the sliding tail stops. As shown here, I coupled the pipes together in pairs and mounted them to plywood sheets using pipe flanges. Then I triangulated the pipes with some 1x4 furring strips for rigidity.

I put the sliding tail stops on the top lengths of pipe and spanned the pipes with a 2x4 crosspiece. A 4x4 spacer atop the crosspiece gave me the distance I needed to reach the ceiling joists and brought the plane of the drywall above the ends of the pipe, allowing me to move the drywall around for final adjustment.

To use the lifts, I lowered the tail stops to about 4 ft. Then I simply placed the drywall onto the crosspieces and pushed up alternately on the tail stops until the drywall reached the ceiling.

—*John D. Leonick, Dorset, Vt.*

DRYWALL KICKER

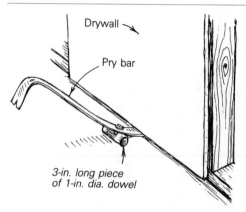

Drywall

Pry bar

*3-in. long piece
of 1-in. dia. dowel*

In the course of my construction projects I have to hang some drywall once in a while, but not often enough to justify investing in a real drywall "kicker"—a lever device made especially for lifting a piece of drywall. Instead, I modified my pry bar as shown above to do the same task. Through its nail-pulling hole I attached a 3-in. length of 1-in. dia. dowel with a bolt that is countersunk in the dowel. Voilà! With just a little toe pressure I can lift a drywall panel 2 in. off the floor with this tool.

—*Andrew Kirk, Independence, Calif.*

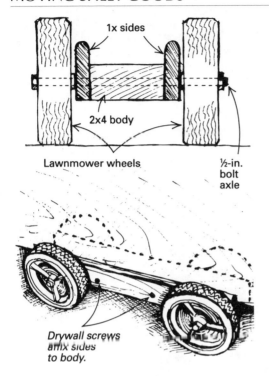

1x sides

2x4 body

Lawnmower wheels

½-in.
bolt
axle

Drywall screws
affix sides
to body.

The little cart shown in the drawing above eliminates much of the toil and the potential damage that accompanies moving heavy sheet goods around the shop or on the job. I've found it to be especially helpful for rolling 12-ft. and 16-ft. sheets of ⅝-in. drywall through the rooms of a house. If its dimensions are increased, the cart can also be used for moving heavy timbers, stacks of floor joists, tubs, water heaters and other heavy appliances.

To make the cart, I started with four 6-in. wheels. I bought mine at a discount building-supply store for about $2 each, where they are typically sold as replacement wheels for lawnmowers (8-in. wheels were also available for $4 apiece). I also bought two ½-in. by 8-in. machine bolts (with nuts) for axles.

I started assembly by tightening a nut up to the unthreaded portion of each bolt. Then I cut off the remaining threaded portions of the bolts with a hacksaw. I put the wheels on the axles, dished side out, and measured the distance between them to determine the width of the cart's body plus its sides. As shown in the drawing, a 2x4 body with 1x sides added up to the right dimension. There should be about ⅛ in. of play between each wheel and its side. I made the body about 20 in. long and attached the sides with glue and a couple of drywall screws. Then I rounded the top edges of the sides with a ¼-in. roundover bit.

After drilling axle holes from both sides of the body with a ½-in. spade bit, I put the wheels on and tightened each nut with a wrench so that it jammed against the unthreaded portion of each bolt's shank. I squirted a drop of oil on each wheel, and the little cart was ready for action.

—Mark White, Birmingham, Ala.

DRYWALL PATCH

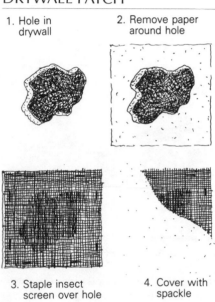

1. Hole in drywall
2. Remove paper around hole
3. Staple insect screen over hole
4. Cover with spackle

Before we moved into our house, somebody ripped out all the wiring—I guess for the copper. Whoever did it punched access holes in the drywall above each outlet box. Rather than going to the expense of installing new drywall throughout the house, I used insect screen and three layers of spackling compound to repair the holes.

As shown here, I first cut through the outer layer of paper and excavated a square or rectangular hole (depending on the shape of the damaged area) about ¹⁄₁₆ in. deep. Then I stapled a piece of insect screen to the drywall and covered the screen with a layer of spackle. I feathered the edges of the next two layers of spackle to blend into the field. The screen keeps the spackle from shrinking as it dries.

—Timothy Carlson, Centerville, Mass.

TEXTURED DRYWALL PATCH

On my current kitchen remodeling job I had to patch a hole in the ceiling and match the existing drywall texture. The old texture looked as if it had been blown on in big globs and then lightly flattened with a trowel. I didn't want to call in my drywall contractor for such a small task, or mask off the cabinets to protect them while I sprayed the 1-sq. ft. patch with topping compound. So I tried another approach.

I picked up a nearly empty tube of latex caulk and squeezed out its remaining contents. Then I put a garden hose over its nozzle end. The water blasted off the tube's plunger and cleaned out the cylinder. Next I filled the tube with drywall compound, replaced the plunger and used my caulking gun to squeeze globs of compound onto the ceiling. Then I flattened the compound with a trowel. Painted to match, the patch is practically invisible.

—Kenneth S. Hayes, Merced, Calif.

DRYWALL BOOSTER SHOT

Recently, an accident left a dent about the size of a tennis ball in our veneer-plaster wall, much like the damage left by doorknobs bumped into walls unprotected by doorstops. The plasterboard underneath was damaged enough that it was too weak to support a plaster patch without cracking again. The gypsum core was shattered but with minimal damage to the paper surfaces. I wanted to repair the existing material without cutting out the damaged piece and did not want to tape and plaster the areas because that would have required a large, feathered patch for a relatively small area of damage. I needed to restore some structural integrity to the core.

I began by drilling a series of holes about ½ in. apart. I used a $^3/_{16}$-in. bit, a diameter that corresponded to the tip of a medical syringe with its needle removed. I drilled holes about half to three-quarters of the way into the core of the drywall and injected each hole with yellow glue until it oozed out of the neighboring holes. After the glue dried, the damaged area became as solid as the original wall.

Syringes can be tough to get if you aren't somehow involved in the medical profession. But most woodworking-supply stores and catalogs have syringes specifically designed for injecting glue. The Woodworker's Store (21801 Industrial Blvd., Rogers, Minn. 55374-9514) is one source.

—David O. Hasek, Stevensville, Md.

GLUING PLASTER

I am restoring an old house. The finish coat of plaster in the front hallway wasn't bonded to the scratch coat in many areas. The traditional fix is to inject white glue between the coats. But white glue takes a long time to dry. I tried something different—cyanoacrylate adhesive (CA). This glue, which is available at hobby shops, is good for gluing porous surfaces. It's strong, wicks well into tight places and sets up fast.

I started my plaster fix by drilling ⅛-in. holes on approximately 8-in. centers. I then squirted the CA into the holes and held pressure on the area for 30 seconds. The finish coat bonded beautifully to the scratch coat. You can get CA in different viscosities. The thin stuff wicks better; the thicker adhesive is better for filling gaps.

—Mark Shilling, Raleigh, N. C.

PLASTER CLAMP

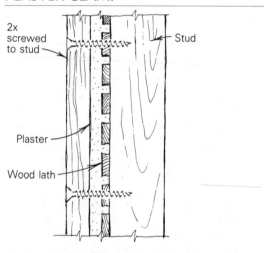

2x screwed to stud

Stud

Plaster

Wood lath

Using a reciprocating saw to cut an opening into a wall made of plaster and wood lath runs the risk of cracking the plaster around the edges of the hole. To minimize the problem, find the nearest stud outside the desired cut line and screw a length of 2x to it, as shown in the drawing above. Now you can use a reciprocating saw alongside the stud to cut the plaster and lath. The 2x will act like a clamp, preventing the vibration from damaging the plaster you want to save.

—Allan R. Aho, Minneapolis, Minn.

DRYWALL EDGE TRIMMER

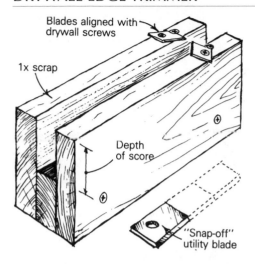

Blades aligned with drywall screws

1x scrap

Depth of score

"Snap-off" utility blade

Whenever possible, our crew uses "in-line" framing for walls to save lumber (weight from above bears directly on studs, and plates are joined with steel splices). Because this style of framing allows a single top plate, our walls are a little less than 8 ft. tall. The obvious downside to this technique is that we end up having to trim about an inch off the ends of our drywall. This is tedious work with a utility knife, so I bought an edge trimmer for $25. It didn't work very well, making the job even more frustrating. Forced to come up with something, I screwed together some pieces of 1x scrap to make a cutting guide and attached a couple of "snap-off" utility knife blades to its top edge (drawing above). When I slide this tool along the edge of a piece of drywall, the blades cut from both sides, making it quick and easy to remove uniform strips from a sheet of drywall.

—*Brian Bush, Dafter, Mich.*

SCRIBE-FIT PLASTER PATCH

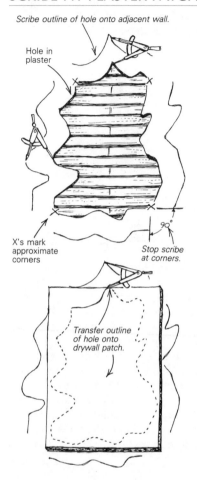

Scribe outline of hole onto adjacent wall.

Hole in
plaster

X's mark
approximate
corners

*Stop scribe
at corners.*

90°

*Transfer outline
of hole onto
drywall patch.*

As a remodeling contractor in Pennsylvania, I often have to patch plaster walls. Unfortunately, "squaring up" a hole in a plaster wall to receive a drywall or rock-lath patch can be mighty frustrating. It seems the more I cut to square the hole, the more plaster falls out. As a result, I've found it easier to cut a piece of drywall to fit the ragged shape of the opening. But the patch has to be a good fit. If there are wide gaps around the edges, the filler tends to shrink and crack. The drawing here shows a scribing method that I've developed to make an accurate patch quickly. Although it looks kind of complicated, the whole operation takes about five minutes.

I begin by removing any loose plaster from the edges of the hole. Then I mark the approximate corners with X's. They serve as registration points for starting and stopping the scribing, and their location isn't critical.

Using a compass or a scribing tool, I scribe lines on the wall that duplicate the four edges of the hole. I usually set the distance between the pointer and the pencil between 3 in. and 4 in. While the setting is arbitrary, it shouldn't be changed once set. As shown in the drawing, the top and bottom edges of the hole are scribed with the compass oriented vertically. The sides are scribed with the compass held horizontally.

Next I cut a piece of drywall that is larger than the hole, but fits within the scribe lines. The one shown here is rectangular, but a rough patch can take on a pretty strange shape all by itself. I press the patch against the wall and, using the scribe marks on the wall, transfer the shape of the hole onto the patch. Now I can trim to the cutout line with a drywall saw and shave the edges as necessary with a utility knife for a perfect fit.

—*Mark Benzel, West Grove, Pa.*

REMOVING DRYWALL

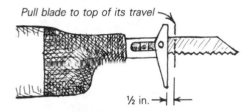

Pull blade to top of its travel

½ in.

If you've ever had to slug it out with a wall to remove drywall for a remodeling project, you can appreciate how tedious it is. Faced with this problem on a recent job, I came up with the following method to cut the drywall into manageable chunks.

As shown above, I affixed an old, general purpose wood-cutting blade to my reciprocating saw (after unplugging it, of course). Then I pulled the saw's drive shaft outward, to the limit of its travel, and marked a cutline on the blade ½ in. from the shoe. This represented the thickness of the drywall I was removing. Next I used a pair of cable pliers to cut the blade.

In use, I simply held the saw against the wall with its shoe flush against the surface. As I moved it up and down the wall, it punched a series of closely spaced holes in the drywall. Studs didn't interfere—the saw just passed right over them. In no time I had a tidy pile of ex-walls, without a cloud of asphyxiating dust.

—*Patrick C. Perry, Lakewood, Colo.*

DRYWALL PATTERN FOR COMPLEX CUTS

Sometimes a piece of drywall has to take on a rather contorted shape to fit a complicated wall or ceiling. Add some cutouts for light fixtures, switches or outlets and you've got a piece that represents a good hunk of time and attention. When I'm faced with laying out such a piece, I use a pattern made of ½-in. rigid foam insulation. This material comes in lightweight 4x8 sheets, so lifting it into place to check alignment is easy. To mark fixtures, I simply push nails through the foam to locate the corners of electric boxes. Then I lay the insulation atop a piece of drywall and tap on the nails to transfer the corner marks to the drywall. If I need to make cutouts in the foam, I save them so that I can use sealing tape to put the foam board back together and use it for insulation.

—*Jim Klahn, Seattle, Wash.*

LIGHT-FIXTURE TEMPLATE FOR DRYWALL

I see more and more recessed light fixtures in ceilings these days, and they present a layout challenge when it comes time to cut the corresponding holes in the drywall. Faced with this job, I put the plastic lid from a 3-lb. coffee can to work. This size lid fits snugly over the rim of most standard recessed housings.

To use the lid, I first push a nail through the center point of the lid. Then I snap the lid over the rim of the housing, and measure from the appropriate framing members to find my center-point coordinates. Then I put the lid on the drywall to be cut, nail at the center point, and trace around it. The layout is complete.

—*Daniel J. Ewing, Gibsonia, Pa.*

REPLACING CEILING TILES

Our company recently remodeled an office suite, and part of the job included replacing the ceiling. The new ceiling was to be of the same material as the old one—acoustical tile. Before starting the job we noted that the ceiling had a lot of complicated cutouts for recessed lights, smoke alarms and air-conditioning ducts. So during the demolition phase of the project, we took down only the full-size tiles, leaving the specials in place. When it came time to put in the new ceiling tiles, the original ones became the templates for all the necessary cuts. This technique saved us a lot of time.

—*Andy Levinson, Navesink, N. J.*

SCRIBING EXTENSION JAMBS

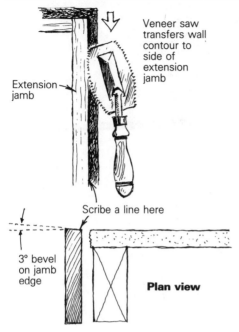

Veneer saw transfers wall contour to side of extension jamb

Extension jamb

Scribe a line here

3° bevel on jamb edge

Plan view

Fitting extension jambs to the irregular walls of older homes can be time consuming. Faced with this task, I get out my veneer saw. First, I rip the extension jambs slightly wider than necessary. Holding a jamb in place with one hand, I slide the saw along the wall so that its teeth scribe a line on the back of the extension jamb, as shown here. A veneer saw's teeth have no set, so they will not mar a finished wall, and the serrated cutting edge is not so apt to wander with the grain the way a knife edge would. Next, I use a plane to trim down to the scribe line at a 3° bevel so that the inside edge of the jamb will stand slightly (and properly) proud of the wall.

—Steve Becker, Valatie, N. Y.

TAPING KNIFE PANS

If you're having difficulty finding a large enough pan for your biggest drywall taping knife, try a piece of vinyl rain gutter fitted with two end caps. In addition to coming in any length you want, a vinyl pan is rustproof and easy to clean.

—Joe Graczyk, Cazadero, Calif.

BOXED LALLY

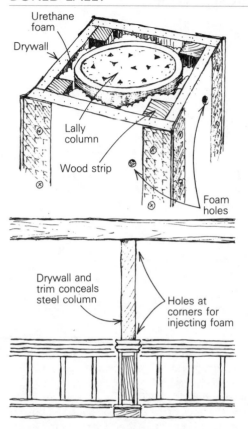

Urethane foam

Drywall

Lally column

Wood strip

Foam holes

Drywall and trim conceals steel column

Holes at corners for injecting foam

While finishing a downstairs room, my thoughts were on the two Lally columns that carry the beam supporting the ceiling. Their steel-pipe look didn't go with the new decor, which featured a chair rail 40 in. above the floor, wainscoting, baseboards and drywall. I decided to apply those details to the columns. Here's how I did it.

First, I boxed the columns with drywall. At the corners of the boxes, I screwed the drywall to straight-grain pieces of pine, ripped into ¾-in. by ¾-in. strips. The pine rippings reinforced the drywall corners from the inside. Then I applied the metal corner bead, again using screws to keep from banging the boxes apart. The problem then became how to secure the drywall boxes to the columns. As shown above, I drilled small holes next to each corner of the boxes at three elevations: near the top; 40 in. above the floor; and at the bottom. Next I poked the nozzle of a urethane foam can through each hole, depositing a generous dose of the expanding foam at each corner. The foam is very sticky, and it pretty much stays where you put it so I didn't have to fill the entire box. A couple of hours later the foam was hard, and the boxes were surprisingly sturdy. Trimmed out with chair rail, wainscot and base, the columns (and the room) look great.

—*James Whidden, Westminster, Mass.*

LALLY BOXING

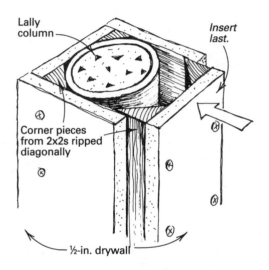

Lally column

Insert last.

Corner pieces from 2x2s ripped diagonally

½-in. drywall

I recently boxed my Lally columns (to meet fire-protection codes rather than for appearance), and I used an approach similar to the tip by James Whidden (facing page). The foamed-in-place method described in that tip is effective, but you can leave out the foam if you rip the corner pieces at 45° (I made mine out of 2x2s). The triangular cross section allows the drywall to fit tightly against the column, saving space and eliminating the need to be anchored.

As shown in the drawing above, I assembled a three-sided box and placed it around the column. Then I slipped the fourth side with its pre-attached corner pieces into place, and screwed it to the adjoining pieces.

—*Chet Burgess, Wappingers Falls, N. Y.*

WING-WALL REINFORCEMENT

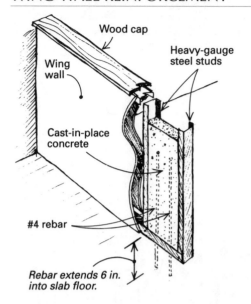

Wood cap

Heavy-gauge steel studs

Wing wall

Cast-in-place concrete

#4 rebar

Rebar extends 6 in. into slab floor.

I worked on a commercial job last year where we had to build an office that included a 42-in. tall, 5-ft. long wing wall. The wall was built out of metal studs, covered with drywall and finished along its top with a wood cap. The wall was held at one end by a bearing wall, but the only anchors holding the rest of the wall were the powder-actuated fasteners driven through the track into the slab. This had me worried. I was sure that, in time, people leaning on the wall would eventually weaken it to the point of collapse. So I strengthened the wall with concrete.

First I drilled a couple of ½-in. dia. holes through the track and 6 in. into the slab for a pair of #4 rebar. As shown here, I placed them in the outermost bay of the wall. I used heavy-gauge steel studs for the two outer studs and turned them so that the flanges faced each other to lock into the concrete better. Then I screwed some scrap ⅝-in. plywood to the two studs, making a 3-ft. tall form. Filled with concrete, the end of the wall became a cantilevered beam that is quite strong. The drywallers glued the drywall to the concrete, and we were done.

—Chris Sturm, Hamburg, Pa.

MOUNTING STRIPS FOR WALLS

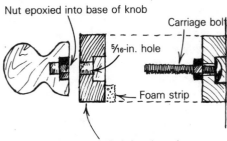

Nut epoxied into base of knob

Carriage bolt

5/16-in. hole

Foam strip

Wood strips accept bolt heads and nuts

Hanging posters, artwork, calendars and fabric wall pieces can be a problem, especially in children's rooms. Tape and adhesives dry out and leave residues. Tacks and nails leave crumbling plaster and holes to patch. Corkboard is a pretty good solution, but to be useful it has to cover a lot of wall, and the accompanying tacks aren't advisable in rooms for young children. To avoid all of these problems, I designed the mounting system shown in the drawing above for use in my son's room.

My mounting strip is a long clamp made up of two ¾-in. by 2-in. strips of wood. The strips are drilled and counterbored for ¼-in. carriage bolts 2½ in. long. The board that applies the pressure has 5/16-in. holes drilled in it to accept the bolts and a strip of foam weatherstripping along the bottom edge to better grip whatever gets placed between the strips. I drilled the bolt holes 21 in. apart for strips 25 in. long. To apply the clamping pressure, I used wood drawer knobs with nuts epoxied into their bases.

—John Roccanova, Ancramdale, N. Y.

CLOSET-SHELF BACKING

Whenever I apply sheathing to closet walls that are to have several shelves, I use ½-in. birch plywood for the end walls. The plywood makes it easy to hang a variety of adjustable shelf systems, such as track or peg-type shelf supports, or permanent shelf cleats without worrying about the location of the studs. The plywood-to-gypboard corner can be taped conventionally or caulked with a good grade of siliconized latex caulk.

—Wayne Breda, Hope, Maine

REMOVING WALLPAPER PASTE

Here's a little trick I stumbled upon that is too sweet to keep to myself. After removing some heavy-duty wallpaper, which came away from the wall easily without the use of a steamer, I was left with a heavy residue of dried glue. It gave the wall a rough texture unsuitable for painting. Rather than try to remove the glue, I decided to smooth the wall with a thin layer of joint compound. After a few minutes of skimming on the compound, I noticed that some areas were bubbling up. When I tried to touch up the spots, I realized the glue had softened. I decided to try to scrape the compound off one area, and to my delight, the compound came off easily along with all of the glue, leaving a clean, smooth surface. Evidently the moisture in the compound softens the glue, while providing a medium to carry the glue off the wall.

I found I could cover about 100 sq. ft. of wall with compound before going back to remove it. This allowed the glue to loosen for about 10 to 15 minutes, which seems ideal. If compound sits too long and won't come off completely, simply wipe it with a wet sponge and let it sit a few more minutes.

I cleaned up with a sponge-washing of trisodium phosphate followed by a rinse of clear water. Wall dried, I applied the finish coat of paint over an alkyd base primer.

Incidentally, the last time I used this technique, I used up three buckets of joint compound that had been frozen, and consequently rendered useless for drywall work.

—Paul Hirsch, Stamford, Conn.

SAWDUST STRIPPING

Reading about Paul Hirsch's discovery that joint compound will remove wallpaper paste (see the facing page) reminded me of the time back in 1960 when I made a similar discovery. I was a struggling jack-of-all-trades down in Florida, and I'd take any job to survive. One dubious endeavor was to strip the paint off a pair of church doors. They were 4 ft. wide, 9 ft. tall and made of 4-in. thick cypress. They were deeply carved on both sides with the faces of saints, and the paint job probably went back to Ponce de Leon.

One night while fighting a deadline, I ran out of steel wool and rags. In desperation I flooded one of the panels with paint stripper and let it soak for a few minutes. Then I spread out a 1-in. thick layer of sawdust and worked it across the panels and into the crevices with a scrub brush. The results were amazing. The sawdust absorbed the paint, speeding the job fourfold.

I use a bristle or brass brush on woodwork or antiques—a steel wire brush will ruin a nice piece of wood. I prefer the runny types of paint stripper (they're cheaper too). If the brush starts to plug up, throw on more sawdust. I've used this method to strip picture frames, fancy baseboards, trim and carved furniture, all with good results.

—Clyde R. Kennedy, Rushville, Ohio

5

DOORS AND WINDOWS

BETTER BACKING FOR DOORS

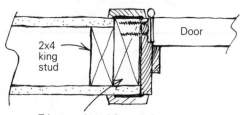

Trimmer ripped from 2x6

If you think about how most doors are mounted to most jambs and visualize the relationship of hinge-to-door and hinge-to-jamb, you may see some room for improvement. Standard butt hinges, heavy-duty or not, have a hole pattern that permits two screws at best to reach the framing trimmer. The thicker the wall finish, the more tenuous is the screw's anchoring ability. When the doors are light, the holding power of the screws will most probably be adequate. But when the hinges on heavy, solid-core doors are secured to such minimal backing, the hinges can pull away from the jamb over time.

My solution to this problem is to rip a trimmer that equals the width of the king stud plus the thickness of the wall finish. As shown in the drawing above, the deep trimmer allows full backing for the hinges. The wall finish abuts the edge of the trimmer, so the casings should be wide enough to conceal the joint between them.

By the way, if you choose to let the trimmer overlap on both sides of the door there are two advantages: the framing is prepared for a door to be hung on either side of the jamb; and the drywallers will be forced to cut their drywall to fit, meaning the drywall won't extend past the trimmer, which makes it a chore to install the shims.

—*M. F. Marti, Ridgway, Colo.*

CHISELING HINGE MORTISES

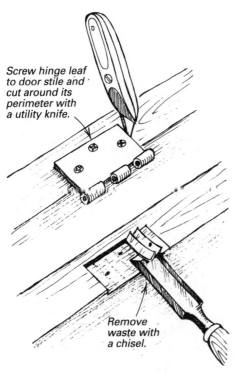

Screw hinge leaf to door stile and cut around its perimeter with a utility knife.

Remove waste with a chisel.

I typically use a router and a template to cut mortises or to enlarge existing mortises for new door hinges. But for small jobs where it doesn't make sense to cart along a lot of gear, I use a utility knife and a chisel for the same purpose. First I screw the hinge to the door stile in the desired position. Then I score around the edges of the hinge with the knife, as shown here. With the hinge removed, I chisel the mortise to the thickness of the hinge. Now I can re-attach the hinge using the same screw holes.

—*Daniel E. Hill III, Griswold, Conn.*

REPLACING A DOOR JAMB

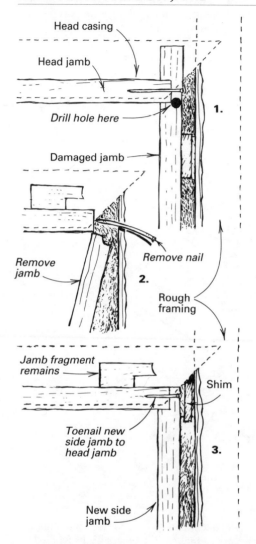

Occasionally I'm hired to replace door jambs that have been damaged by burglars. Often, only one jamb leg is damaged (the strike side of the door). The other side jamb and head jamb remain serviceable, so replacing the strike jamb without messing up the rest of the installation is the problem. Here's how I do it.

I begin by adding some extra finish nails to the door casings on the good leg and the top jamb. This step helps to hold everything in place. Then I pull the casings from the damaged jamb. Drawing 1 above shows the point at which I drill a ¼-in. or ⁵⁄₁₆-in. hole in the side jamb next to the head jamb. This hole lops off the top of the damaged jamb. To remove the jamb, I cut it in half at about the height of the strike plate and pry away the two pieces.

Drawing 2 shows the jamb being removed and the fragment of the jamb top lying on the head jamb. If there's room I pull this scrap out. If not, there's usually room to shove it out of the way up top. If I can reach them, I pull the nails out of the head jamb with needlenose pliers. If I can't, I use a reciprocating saw to nip them off.

Drawing 3 shows the new jamb in place. Note how I've cut off the end of the new jamb, giving it a rabbet instead of a dado. This allows the new jamb to be tucked into place if the clearance between the rough framing and the head jamb is tight. If there's room I leave the dado in place. In either case, I toenail the new side jamb to the head jamb and back the joint with shims to keep it tight. Now the new jamb can be plumbed with shims or adjusted to match existing conditions.

—Chris Dahle, Denver, Colo.

REFITTING A DOOR

As a house ages, the openings that used to be square sometimes settle into parallelograms, and doors that used to close don't anymore. I repair a lot of old doors, and in my travels I've seen many attempts to make a formerly square door fit into a tilted opening.

Usually a home owner will remove material from the door stile that is overlapping the jamb. If there's just a little overlap, this method will work. But if the overlap is substantial, things can get complicated in a hurry, and the amount of time required to remedy the situation grows accordingly. Locksets need to be rebored, and sometimes the holes need to be patched because the door's handle trim isn't wide enough to conceal the gaps. An equally complicated approach is to plane away material from the door jamb. This requires moving door strikes and weatherstripping, and it can throw off the spacing of the casing reveals around the door.

The simple way to treat this problem is to remove material from the door's hinge stile. Measure the door overlap and add $\frac{1}{16}$ in. Mark these measurements on the hinge stile and remove the door for trimming. Now cut new mortises for the hinges, using the old screw holes to position them, and you're ready to rehang the door.

—Theodore F. Haendel, Great Neck, N. Y.

POCKET-DOOR FIX

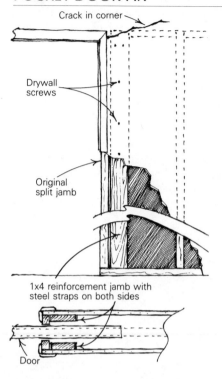

Crack in corner

Drywall
screws

Original
split jamb

1x4 reinforcement jamb with
steel straps on both sides

Door

Soon after reading Kevin Ireton's article on pocket doors (FHB #54, pp. 63-66), I needed to fix a pocket door in a 2x4 wall. Its skimpy ¾-in. split jamb had flexed at the junction between the split jamb and the header. Result: cracks in the drywall at the corners. To solve this, I stiffened the walls near the split jamb without thickening them or tearing them open.

First I removed the door stops and the door, allowing access to the inside of the wall. Next I cut a couple of 1x4s to fit between the header and the plate adjacent to each half of the split jamb, as shown in the drawing above. The 1x4s are meant to take a load from the side, reducing the bending that caused the cracks. To make them extra stiff, I screwed ⅛-in. thick by ¾-in. wide steel straps to their edges with drywall screws on 1-ft. centers.

Before slipping each 1x4 in place, I ran a liberal bead of drywall adhesive down the face that bears against the back of the drywall. Then I used 1⅛-in. drywall screws to secure the 1x4s to the finished walls. After patching the cracks and holes, the walls were ready to paint.

—*Bruce G. Koprucki, Chaska, Minn.*

DOOR JIG

Stop blocks
align door
with wall plane

Shim jamb to sides of door

When I have a batch of interior doors and jambs to install, I use a jig that I made out of a cheap hollow-core door. I paid $10 for the 2-ft. 6-in. by 6-ft. 8-in. door, and then removed most of it, leaving about 6 in. of door around its perimeter, as shown above. The door was pretty flimsy at this point, so I reinforced it between the faces with 1x plywood around its cut edges. Small wood blocks screwed to the corners align the jig with the plane of the wall.

In use, I shim the jamb sides snug to the sides of the door. The opening in the door jig makes it easy to step from one side to the other as I shim the jamb. A perfectly sized and plumb jamb takes less than five minutes to do. When I get ready to hang a door in its jamb, I reduce its width by about ⅜ in. as I bevel the door's edge.

—Steve Prince, Los Alamitos, Calif.

INSTALLING PRE-HUNG DOORS

I'm a lazy perfectionist, which I think helps to explain the origins of the following method that I use to install the jambs of pre-hung doors and the extension jambs and sills of windows. Once I tell you about the door technique, I'm sure you'll be able to imagine how to apply it to the installation of windows.

I begin by checking the hinge-side cripple stud for plumb. If it needs wood shims at the top or bottom, I add them as necessary. I then use my finish nailer to affix the hinge-side of the door jamb to the cripple at the top and bottom. Next, I sink a #2 by 10 self-tapping Phillips-head screw beside each hinge gain. I place the screws to end up under the door stop, and I back them with shims if necessary.

Now comes the easy part. With my nailer in one hand and my pry bar in the other, I sink a few pairs of 8d finish nails through the head jamb and into the header. Usually, the head jamb gets pushed up a little too far in places, so I pry it back down, making the space between the door and the head jamb just right. I do the same thing for the latch-side jamb, nailing and then prying the jamb back to get even spacing from the door. This works because with pneumatic finish nailers, you can tack the jamb in place without a shingle backing the nail.

With the door hung in its rough opening and the jambs properly aligned, I use a can of urethane foam insulation to fill the crevices between the jambs and the rough framing around all the nails. The foam sticks tenaciously, and provides a firm, non-shrinking backing in an instant. Later on I use a loose hacksaw blade to trim off the cured, excess foam before I nail on the casings.

—Armin Rudd, Cocoa, Fla.

DOOR-HANGING HELP

Whenever I have to hang a door—especially a heavy one—I get an assist from a 1-ft. long piece of ½-in. EMT (electrical metallic tubing). A piece of this smooth, sturdy metal conduit makes a good roller when it's placed under the door, allowing me to roll the door up to the hinged door jamb. Then I can use the EMT as a fulcrum to pivot the door up to hinge height. This trick has saved me a toe or two, plus my back.

—Alan Zebker, Santa Monica, Calif.

SOLO LEVEL SHOTS

A while back I had to install interior doors in the basement of a custom home. My boss instructed me to make sure all the jamb heads were the same height so that all the trim would eventually line up. He left me a builders level (similar to a transit but without the up-and-down feature) to check my readings. Then he disappeared.

I got the level set up quickly, then pondered my next move. I wanted a common level line on each side of all the door openings. But while peering through the eyepiece of the instrument, the problem of marking the level lines while standing 20 ft. from the target occurred to me. What to do? Run upstairs and drag somebody away from their work? Get the boss? There had to be a better way.

I spotted a discarded newspaper in the corner of the basement, and suddenly a light went off. I tore strips of newsprint about 3 in. wide and 1 ft. long and stapled one on each side of the door openings in the vicinity of the jamb heads.

I walked back to the instrument, squinted through the eyepiece and focused on a door jamb. I could easily read whatever word was bisected (or underlined) by the level line. I read both sides of a doorway to save time, then walked over and made my pencil marks by the appropriate words. I made a quick check to ensure my accuracy, then moved on to the next doorway. I was able to do several doorways in just a couple of minutes by myself.

I've since repeated this procedure at a couple of other job sites, and I always make sure to leave the newsprint stapled up as long as possible. It drives other tradesmen nuts while they try to figure out why it's there.

—Lloyd Dorsey, Wilson, Wyo.

PLUMB-BOB ANCHOR

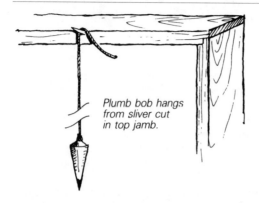

Plumb bob hangs from sliver cut in top jamb.

Whenever I'm hanging a door with wood jambs, I use a chisel or utility knife to lift a small sliver in the center of the head jamb. Then I use it to anchor the string on my plumb line, as shown in the drawing above. The sliver is a quick, handy way to get a purchase for the plumb line, and it stays there until it is eventually hidden behind the door casing. The same trick can be used to hang a plumb line when you're framing.

—Michael R. Sweem, Palo Alto, Calif.

CARRYING DOORS

One day when I was an apprentice, I was complaining about having to hand-carry some solid-core doors down a long corridor. The superintendent, an old carpenter, looked at me with an air of disbelief. "Don't tell me you call yourself a carpenter when you don't even know how to carry a door," he said.

Then he put his back against one of the doors, bent his knees just a little bit, and with his arms stretched down, he grabbed the door, leaned forward and walked away with the door more or less on his back. Ever since, that's the way I carry heavy doors, and so does everyone else to whom I show the trick.

—Robert Doucet, Montreal, Que.

MAKING A QUICK BUCK

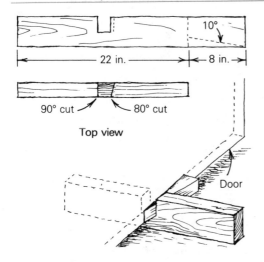

An Irish craftsman, with whom I had the pleasure of working, showed me how to make a simple door buck quickly. In fact, it's so easy to make that it could be considered disposable, but it's small enough that you might want to hang on to it.

Take a 30-in. 2x4 and make a 10° taper cut along the first 8 in. of the face, as shown above. Then make a cross cut at the 8-in. mark so that you have an 8-in. wedge with a 10° taper. Turn the 2x4 on edge and set your saw at a depth of 2 in. Make a 90° crosscut near the center of the board, then move over 2 in. and make another cross-cut at 80°. Then make several passes with the saw between the two cuts, knock out the pieces and clean up the slot with a chisel.

Slide a door into the slot in the 2x4 and drive in the wedge with a few taps. To keep the door from rocking, locate the buck toward one end of the door, or use one at each end.

—*Tim Hoisington, Greenfield, Mass.*

ANOTHER QUICK BUCK

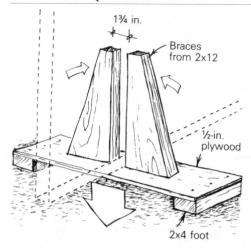

When I have to brace a door for hinge or lockset operations, I use a door buck made from scrap materials like the one in the drawing above. The base is made of ½-in. plywood and a couple of 2x4 offcuts. It's important that the plywood be thin enough to bend a little under the weight of the door, which makes the 2x braces angle inward, pinching the door and holding it steady. If you're concerned about marring the finish of a door, make the gap between the braces a little wider and line the braces with strips of carpet.

—Ed Wilson, Seattle, Wash.

BORING METAL-SKINNED DOORS

An article on installing locksets *(FHB* #79, pp. 40-45) omitted one problem that I frequently encounter: the metal-skinned door. The problem usually crops up when I have to install a deadbolt, and I've neglected to tell my supplier that I need a double-bored door. When this happens, I reach for my adjustable bit and my brace. The threaded center of the bit draws it into the door with just a dimple from a nail or an awl as a starter. Then the sharp spur of the bit cuts the metal just like a can opener. This can be a bit tricky because it takes a little tilting to get the spur through the metal without the chisel portion of the bit engaging the metal.

Once the circle of metal has been cut loose, I back the bit out and remove the disk. Then I continue boring with the brace and the bit, making sure that I've cut the metal circle on the other side before finishing the hole.

—Robert Countryman, Ranger, Ga.

DRILLING FOAM-CORE STEEL DOORS

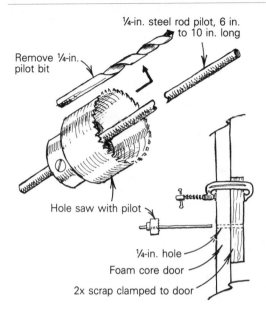

¼-in. steel rod pilot, 6 in.
to 10 in. long

Remove ¼-in.
pilot bit

Hole saw with pilot

¼-in. hole

Foam core door

2x scrap clamped to door

I had to drill holes for deadbolt locks in some steel doors that had foam cores, and I almost destroyed the first couple of doors using a standard hole saw because the ¼-in. pilot bit would wander off course in the soft foam. To remedy the situation, I removed the pilot bit and replaced it with a piece of ¼-in. mild-steel rod, as shown in the drawing above. To use the modified saw, I first drilled a ¼-in. pilot hole through the door and a 2x block clamped to the door. Using the pilot hole to guide the steel rod, I cut my holes with no wobble and no tearout. Incidentally, for this work I found that a bi-metal hole saw held its edge better than one made of steel.

—E. F. Bott Jr., Troy, N. Y.

DRILLING FOAM-CORE DOORS

Like E. F. Bott (above) I had to drill a few holes in foam-core steel doors for locksets. But not wanting to dull my hole saw, I used the following technique. I drilled through the door using a bit the same diameter as the pilot bit in my hole saw. Then I inserted the pilot bit in the hole, and turned the hole saw by hand to mark the cutout on both sides of the door. Using the drilled hole as my starting point, I cut along the line with a fine-tooth metal-cutting blade mounted in my sabersaw, which I adjusted to the scroll setting. The cut was clean, and my hole saw stayed sharp for wooden doors.

—Bruce Weik, Portland, Maine

LOOSE SCREWS IN STEEL DOORS

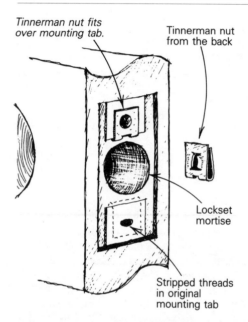

Tinnerman nut fits over mounting tab.

Tinnerman nut from the back

Lockset mortise

Stripped threads in original mounting tab

After years of opening and closing my steel entry doors, I found that the screws holding the latch bolt to the edge of the door stile had started to loosen. The threads in the little mounting tabs that accept the screws had stripped out, leaving oblong holes that afforded no purchase for the screws. Fortunately, I found a simple way to fix the problem.

At my local hardware store, I bought a dozen "tinnerman nuts." As shown above, this kind of nut is made of a piece of folded sheet metal. I slipped the nuts over the damaged mounting tabs in my doors, and now the screws stay put.

—Stephen N. Denton, Somerville, Tenn.

LOCATING LATCHES

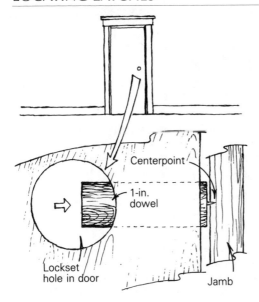

Here's a fast, accurate method that I use for locating the centerpoints for door latch and deadbolt mortises. First I center a hole for a 6d nail in the end of a short piece of dowel (a 1-in. dia. dowel for a typical lockset). Then I drive the nail partway into the hole, cut it off and sharpen the remaining shank to a point.

After boring the holes for the lockset in the door's stile, I insert the dowel into the hole that will contain the latch tube, as shown in the drawing above. Then I close the door, and simply push the dowel marker toward the jamb. The sharpened nail marks the exact centerpoint for my latch or bolt hole.

—Bernard H. King, Mechanicville, N. Y.

DOOR-PULL JIG

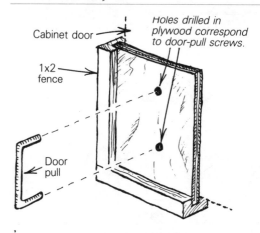

Cabinet door

Holes drilled in plywood correspond to door-pull screws.

1x2 fence

Door pull

Whenever I need to install pulls on cabinet doors, I begin the job by making a jig to locate the screw holes. The jig (drawing above) is a piece of thin plywood (¼ in. to ½ in. works fine) bordered on two sides by 1x2 fences that meet at 90°. Grooves in the 1x2s accept the plywood insert, creating a fence on both sides of the jig. Once I have decided where I want the pulls to be in relation to the corner of the cabinet doors, I drill corresponding holes in the plywood insert, as shown.

To use the jig, simply snug the fences against the corner of the cabinet door where you want to install the pull, and drill your holes using those in the plywood as a guide. For the adjacent door, flop the jig and you're ready to drill.

—*Andrew George, Richmond, Va.*

REPLACING GARAGE DOOR PANELS

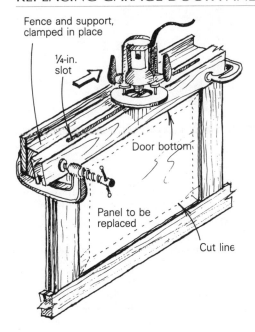

Fence and support, clamped in place

¼-in. slot

Door bottom

Panel to be replaced

Cut line

My frame-and-panel garage door had a nasty gouge in one of the bottom panels. My job was to figure out a way to remove the ¼-in. thick panel without ruining the rest of the door. To do so, I turned to my router.

After removing the door and placing the damaged panel upside-down, I clamped a 2x support adjacent to the bottom frame, as shown above. Next I used a plunge router fitted with a ¼-in. straight-flute bit to make a slot in the door bottom. I had to make the slot about 1 in. deep to get to the panel, and I made my passes in ¼-in. deep increments to avoid overloading the router or straining the bit. Slot cut, I ran my jigsaw around the perimeter of the damaged panel, cutting it into easily removable pieces.

To finish the job, I slid the new panel in place through the slot in the bottom of the door. Then I filled the slot with a glued-in-place piece of ¼-in. stock.

—Elliot Eisenberg, Conyngham, Pa.

TURBINE DOOR

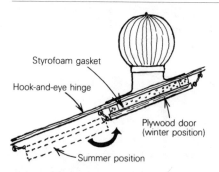

Styrofoam gasket

Hook-and-eye hinge

Plywood door
(winter position)

Summer position

I installed wind turbines on the roof of my house to minimize heat
buildup during the summer. But in the winter, I got tired of going
up on the roof to cover the turbines to prevent rapid heat loss.
Since I have ready access to the attic of my house, I came up with
the hinged door shown in the drawing above to seal off the
turbines in the winter.

The plywood doors are secured to the underside of the roof
decking with hook-and-eye hardware, which allows them to pivot.
When it gets cold outside, I pivot the doors to the closed position,
where they bear against a Styrofoam gasket to cut down on air
infiltration.

—V. A. Maletic, Antioch, Calif.

SHELTERED WOODSHED

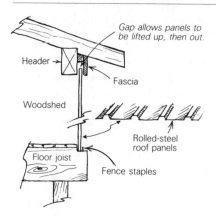

Gap allows panels to
be lifted up, then out.

Header

Fascia

Woodshed

Rolled-steel
roof panels

Floor joist

Fence staples

Our woodshed is near the driveway, where it is easy to load directly
from our pickup. Unfortunately its loading bay faces into the
weather, so I had to add a door to protect our firewood from rain

and snow. As shown in the drawing on the facing page, I used some leftover steel roofing panels that match our roof to make a removable wall.

At the top, the panels are held captive by the header and the fascia. At the bottom they are secured by the threshold on the inside and a row of fence staples on the outside. A space at the top of the panels gives me enough room to lift them out when I need to open the shed.

I've found that the 7-ft. high by 2-ft. wide roofing panels are sturdy enough to stand up to a strong wind. Lapped in the intended manner, they haven't leaked, and the structure looks right at home next to our house.

—Ron Milner, Grass Valley, Calif.

SLOWING DOWN THE COLD

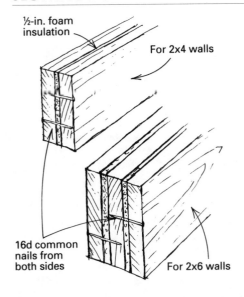

½-in. foam insulation

For 2x4 walls

16d common nails from both sides

For 2x6 walls

When nailing together built-up window and door headers, I use ½-in. foam insulation between the 2x stock instead of ½-in. plywood (drawing above). The sheathing—whether foam board or plywood—is acting only as a packing material to adjust the thickness of the header, so it makes sense to use something with a high R-value.

—Todd S. Pettinger, Springwater, N. Y.

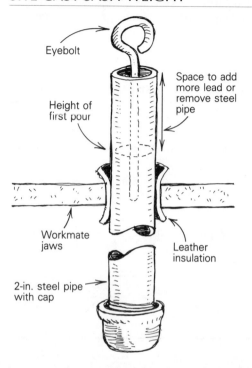

Eyebolt

Height of
first pour

Space to add
more lead or
remove steel
pipe

Workmate
jaws

Leather
insulation

2-in. steel pipe
with cap

I had to restore an old double-hung window that measured 4 ft. by
6 ft., but only one of the two sash weights remained in its pocket.
To make the window operational again, I'd have to come up with
another sash weight. Not having a salvage yard nearby, I resolved
to make my own weight.

I had a 2-ft. length of 2-in. pipe on hand. As shown in the drawing
above, I capped one end of it and clamped it upright in the jaws of
my Workmate®. After figuring the difference in weight between the
pipe and the original sash weight, I got out some old lead pipe to
make up the difference. Wearing gloves, goggles and a respirator, I
melted the lead in a plumber's crucible and carefully poured it into
my steel pipe. I did this outdoors, and protected the area under the
pipe with some firebrick to prevent any spills from causing a fire.
Once the lead was within a few inches from the top of the pipe, I
inserted a 6-in. eyebolt and held it steady until the molten lead
cooled enough to set up. The two weights ended up weighing
virtually the same amount, but if I'd needed to adjust the new one, I
could have added a little more lead to the space at the top of the
pipe, or removed some of the pipe to trim it down a bit.

—*Dave Marlow, Warrenton, Va.*

WORKABLE WINDOW PUTTY

I have made or repaired a lot of window sash, and while I'm pretty good with a putty knife, I still find that glazing takes an inordinate amount of time. One of the minor frustrations is that every can of glazing compound has a slightly different consistency, while there is only one consistency that makes puttying easy, fast and even remotely enjoyable. Even within cans of glazing compounds there are differences between the stuff under the lid and the stuff on the bottom.

I used to add a little bit of paint thinner and mix it in by kneading, but now I've changed to adding just a few drops of Penetrol (H. P. Flood Co, P. O. Box 399, Hudson, Ohio 44236; 800-321-3444), an oil-based paint additive. This seems to make glazing compound much more plastic and adhesive, not just softer. Thumbing on a bedding bead is a breeze, and I can load great gobs of putty on the knife and apply it in one long stroke for my exterior bead. Remember though, only a few drops to a fist-sized ball of compound.

—Daniel Wing, Corinth, Vt.

SEALING AROUND WINDOWS

When I was remodeling a house built in the 1920s, I made the decision to replace all of the leaky, double-hung sash with new windows. My new aluminum windows came with 1½-in. wide flanges that, in new construction, are covered by the siding. In remodeling, however, the flange is nailed on top of, or next to, the old siding, and it has to be covered by trim. However, the trim isn't enough to solve the problem of leakage around the flange. To make matters worse, the building that we were working on had deeply relieved triple-lap siding. A piece of trim nailed to the old siding simply sat atop the high points, leaving enormous gaps that invited water damage.

One possible solution would have been to stuff caulking into all the gaps, but my carpenter had a better idea. He attached 2-in. wide strips of Peel-'n-Seal (Hardcast Inc., 8242 Moberly Lane, Dallas, Tex. 75227) all around the windows, covering both the flange and the old siding. Peel-'n-Seal, which comes in rolls that range in width from 2 in. to 12 in., is a strip of aluminum covered by a ⅛-in. thick layer of asphalt adhesive. It can be easily tooled to conform to any surface, and it will seal wide gaps. We finished our installation with wood trim, entirely disguising our solution.

—J. A. DeCecco, Santa Cruz, Calif.

STRETCHING SCREENS

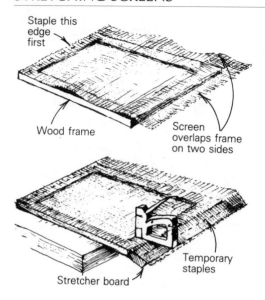

Staple this edge first

Wood frame

Screen overlaps frame on two sides

Temporary staples

Stretcher board

It can be more than difficult to stretch screening evenly across a frame without bags, sags and zigzags. But I've found that with the help of a stretcher board, I can get professional results every time. As shown here, I run the screening a few inches long in both directions. After stapling the screen to one of the short sides of the frame, I staple the other edge of the screen to a stretcher board. After stapling the screen to it, I hang the end of the frame that I'm working on over the end of a table and press down on the stretcher board. This tensions the screen, and with my free hand I staple the screen to the frame. After removing the staples from the stretcher board, I repeat the process on the unstapled sides of the frame. The screen comes out straight and taut.

—David Tousain, Coon Rapids, Iowa

CLERESTORY WINDOW OPENER

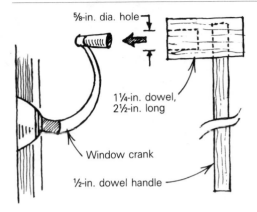

⅝-in. dia. hole

1¼-in. dowel, 2½-in. long

Window crank

½-in. dowel handle

Our solar house has nine clerestory windows, and four of them can be opened for ventilation. But opening them requires either a stepladder, a motorized opener or a 7½-ft. tall assistant.

After pondering the problem, I made my own long-handled opener from a 4-ft. length of ½-in. dowel and a short piece of 1¼-in. dowel drilled and glued onto its end. I drilled a ⅝-in. dia. hole into the end of the larger dowel to accommodate the window crank The device looks like a corncob pipe with a long stem (see the drawing above).

The opener worked the first time out, and it's still in use after five years. I can even use it in the dark, when the wind blows up in the middle of the night.

—*Bill Stuble, Green River, Wyo.*

ARCHED WINDOW PREPARATION

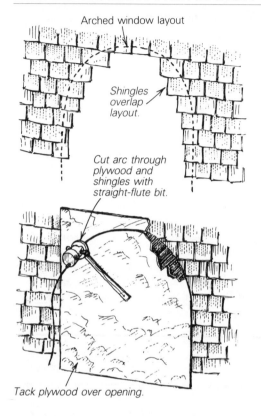

Arched window layout

Shingles overlap layout.

Cut arc through plywood and shingles with straight-flute bit.

Tack plywood over opening.

If I'm installing half-round windows or arched windows with a consistent radius, I take care of the wall finish first. Although this method works with all sorts of siding, let's use shingles as the example. Instead of jigsawing each shingle to fit, I let the shingles overlap the rough opening. While doing this I make sure that no nails are driven within a few inches of the rough opening. This ensures an unobstructed space under the shingles for my flashing.

Next I tack a piece of thin plywood (¼-in. plywood will do) over the area to be occupied by the window. This is my flat work surface. Then I attach a trammel to my router, and I screw it to the center point of the arch. Using a straight-flute bit, I make my first router pass deep enough to cut through the plywood. Then I continue with more passes until I've cut through the shingles. If the pivot point is placed correctly, the opening I get will fit my window and its trim.

—David Hornstein, Arlington, Mass.

6

ROOFS

DAMPENING DRIPS

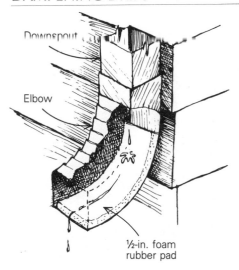

Downspout

Elbow

½-in. foam
rubber pad

After a night-time rain, the residual runoff would slowly
accumulate along the edge of our roof and eventually drain into
the downspout adjacent to our bedroom window. This didn't
amount to very much water, but its intermittent flow would
eventually hit the elbow at the bottom of a downspout with a
resounding drip, Drip, DRIP. This water torture would echo through
our bedroom for hours.

Before resorting to moving the downspout, I decided to try dampening the drips where they landed. As shown in the drawing on p. 103, I used silicone caulk to affix a patch of ½-in. thick foam rubber to the inside of the elbow. I'm relieved to report that peace has been restored to the post-rain portions of the night.

—Richard H. Dorn, Oelwein, Iowa

SLATE ROOF REPAIR

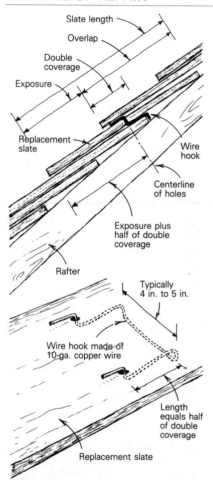

This drawing shows my method for replacing a broken roof slate. Unlike the traditional repair method that is usually performed in our part of the country, this technique leaves no wire exposed to view or to the weather.

After removing the broken slate and finding a suitable replacement, I drill a pair of holes in the new slate. As shown on the facing page, the distance from the bottom edge of the slate to the centerline of the holes is equal to the exposure of the slate plus half its double coverage (that portion of the course where three slates overlap one another). The bent ends of a 10-ga. copper-wire hook pass through the holes. The rest of the wire hook passes under the replacement slate and hangs on the top edge of the slate directly beneath it.

—Randy E. Medlin, Laurinburg, N. C.

SHINGLE PACK

This illustration shows a backpack rig I made for carrying asphalt shingles up a ladder. I started with an old aluminum-framed backpack and removed its bag. Then I added a plywood box to cradle the bundles of shingles. Ropes pass through holes drilled in the plywood box, lashing the box to the frame. About three-quarters of the way up the box, I placed a strap to steady the bundles during transport. Using this rig made it almost fun to carry the shingles up to the roof.

—Henry E. Weis, Kinnelon, N. J.

CLAMP LIFT

For lifting light but bulky items to the roof, such as flashings, I use a locking C-clamp. I welded the ring from an eyebolt to the adjusting screw on the handle of the clamp, allowing me to attach the clamp to a rope easily.

—Derk Akerson, Sacramento, Calif.

SHINGLE CLEANUP

Whenever we have to strip a roof—always a horrible task—we start by spreading a 4-mil polyethylene sheet over the ground near the house. Then we back the pickup as close as possible to the house and build a plywood chute that leads from the truck bed to the edge of the roof. Once we've loaded the shingles in the truck, we drive off the plastic and roll it up, encapsulating any wayward roof debris. We then unroll the plastic at the dump to clean it, and save it for the next job. This method has allowed us to offer our customers the "Dixie Cup Guarantee": the 3-oz. paper cup we give them at the end of the job cannot be filled with what's left on their lawn from the roof.

—Roger Gwinnup, Oxford, Iowa

HEAVY-DUTY STIRRING ROD

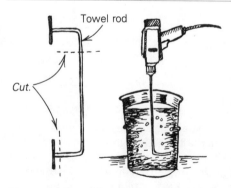

Towel rod

Cut.

After trying several unsuccessful methods to stir asphalt driveway and roof coatings, I happened upon a simple solution. I bought an inexpensive chrome towel bar (mine cost 99¢) and used a hacksaw to remove both of its mounting plates, as shown above. Inserted in my ½-in. drill, the rod became a perfect stirring tool for homogenizing even the most stratified asphalt coatings. As a bonus, its chrome coating allows the rod to be easily cleaned and reused.

—John Luck, Vienna, Va.

DIAL-A-REPTILE

I work for a small architectural office in Boston that does work along the seashore. We sometimes design buildings with low-pitched roofs. When we do, we specify elastomeric membrane roofing because of its good performance overall. It comes in big black rubbery sheets.

Recently we were surprised to see a small rectangular hole in one of our newly installed roofs. It was located next to a seam in the roof where two pieces of membrane had been glued together. We called a few manufacturer's reps, figuring it was some kind of a defect, and found out much to our surprise that it was just another seagull attack.

As it turns out, seagulls are attracted to newly installed membrane roofs. Perhaps the black color and elastic consistency reminds them of whale or fish skin. The gulls like to pick at the seams, especially where the softer flashing sheets overlap the field sheets. After about four or five days, the flashing material cures to its final hardness and the gulls leave it alone.

If you are planning to install an elastomeric roof, take advantage of the seagulls' phobias. They hate snakes. Synthetic rubber snakes are available at most joke shops. Even better are the big inflatable jobs (try the Nature Company, P.O. Box 2310, Berkeley, Calif. 94702). You have to weight them down, but they look pretty lifelike wiggling in the breeze. If you can't get a snake, we were told that an owl might also work.

—William F. Roslansky, Cambridge, Mass.

STAGING PLANK SUPPORTS

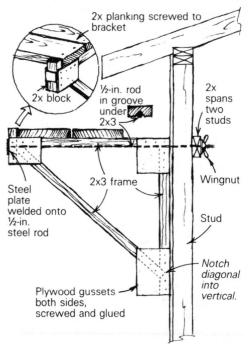

2x planking screwed to bracket

2x block

½-in. rod in groove under 2x3

2x spans two studs

Wingnut

Stud

Steel plate welded onto ½-in. steel rod

2x3 frame

Plywood gussets both sides, screwed and glued

Notch diagonal into vertical.

Whenever my crew and I need to work around the edge of a roof, we erect a row of planks around the roof's perimeter and support it with large brackets that bear against the walls. The brackets are triangular frames made of 2x3s (drawing above) and have plywood gussets at the corners. A length of ½-in. dia. steel rod is let into a groove in the bottom of the horizontal 2x3. It has a welded plate on the outboard end, and it's threaded on the inboard end.

To use the brackets, we first drill a hole in the sheathing at the desired height on the wall, right next to a stud. The steel rod extends through the hole, and then through a scrap of framing lumber long enough to span two studs. The bracket is then secured with a big wingnut. We space the brackets approximately 10 ft. o. c.

I think the staging makes rafter and truss installation much safer and easier, and it seems to speed up eave and soffit detailing, as well as roof-edge shingling. The staging is completely independent of uneven backfill or other ground conditions that bedevil scaffold installations.

—*Peter Evans, Sackville, N. B.*

THE FLASHBOB

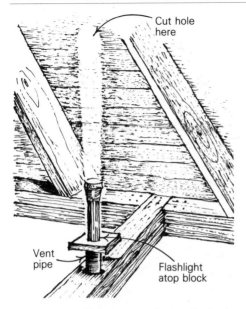

Cut hole here

Vent pipe

Flashlight atop block

As a plumber, I have to cut plenty of holes in roofs for vent pipes. One way to locate the spot for the hole is to hold a plumb bob over the vent pipe stub and then mark the vent's centerline on the roof sheathing. But this can be cumbersome in a tight spot. I've found that a better way to locate my vent cuts is with a narrow-beam flashlight. As shown in the drawing, I place the flashlight atop a wood block over the vent stub. If I'm concerned about the flashlight shining straight up, I check it for plumb with a torpedo level. Held plumb, the light shows the way to my target.

—F. X. Lowry, Somerville, Mass.

REMOVABLE RAIL

I am often asked to build removable duckboards for decks over living spaces so that the roof surface underneath can be cleaned and repaired. Recently, I had to build a deck whose railing was also removable. I built it in 8-ft. sections with 2x4 rails and connected them with hardware made to hold truck stake-sides together. These do the trick nicely, and are weather-resistant and inexpensive.

—Terry Turney, Sandpoint, Idaho

STRAP CLAMP FOR FLUES

Tightening plumber's tape strap-clamp
disengages clips linking flue sections

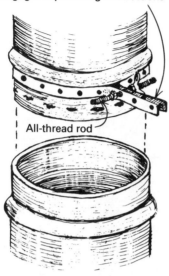

All-thread rod

My client wanted his prefab fireplace lifted to accommodate a
raised hearth. The primary obstacle to this seemingly simple desire
was the 40 ft. of interlocking metal chimney above the fireplace. I'd
have to remove the sections from the top down to separate the flue
from the fireplace before the hearth-raising could take place. I
called some chimney specialists to find out how to separate the
interlocking flue sections. Their advice was to use three
screwdrivers simultaneously to pry the interlocking tabs loose, and
they'd be happy to do the job for about $1,000.

Stunned by this revelation, I mentioned my predicament to my
local sheet-metal supplier. He let me in on a trade secret that works
for both assembling and disassembling interlocking metal flue
sections. Instead of using three screwdrivers, his method involves a
single length of metal plumber's tape, a piece of ¼-in all-thread rod
and a wingnut. As shown above, the plumber's tape goes around
the flue pipe that fits inside the mating section.

Bend tabs on the ends of the tape and leave an inch or so
between them. Insert the threaded rod, put a nut on one end and a
washer and a wingnut on the other. As you tighten the wingnut, the
flue is constricted enough to pop free of the adjacent section.

—*David Strawderman, Los Angeles, Calif.*

7

TRIM
AND SIDING

CUT-LINE REMINDER

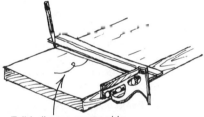

Tail indicates waste side

One-eighth of an inch isn't very big, unless it's the gap between two
pieces of fancy crown molding, caused by cuts made on the wrong
side of the cut line. It's a fundamental mistake, and even seasoned
carpenters make it once in a while. My father was a trim carpenter,
and to avoid this problem he always put a little tail on the waste
side of the cut line, as shown in the drawing above.

—R. E. Stallings, Carrollton, Ga.

MITERING RUBBER COVES

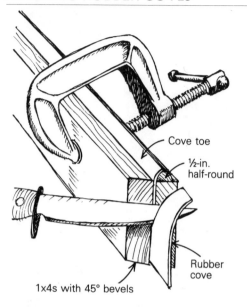

Cove toe

½-in.
half-round

Rubber
cove

1x4s with 45° bevels

Whenever I have to cut a mitered edge in a piece of rubber cove base, I reach for a couple of 1x4s and a C-clamp. As shown above, the 1x4s have 45° angles on their ends. Clamped together with the cove in the middle, they make an accurate cutting guide for a knife blade. To back the cut at the curved "toe" of the cove, I glued a length of ½-in. half-round to one of the 1x4s.

—John Molnar, Moorestown, N. J.

SANDING TABLE

A finish carpenter doing meticulous, stain-grade trimwork can make good use of a stationary sander. It's just the tool for taking off minute amounts of end grain in the quest for the perfect fit. Unfortunately, a stationary sander is too cumbersome to lug around the job site, so we've cooked up a dandy little substitute. As shown here, we've made a cradle that holds a belt sander on its side.

The bottom of the drawing shows the two vertical blocks that surround the handle and the fan/motor housing. A notch in one of the blocks accommodates the handle. We carefully fit the blocks, so we don't even need a clamp to hold the sander in place during operation.

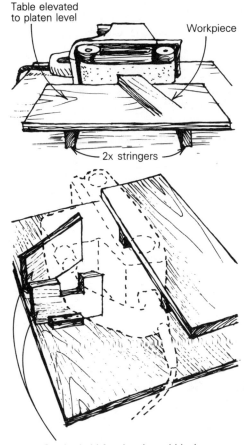

Table elevated
to platen level

Workpiece

2x stringers

Sander held fast by shaped blocks.

The upper part of the drawing shows the sanding table, which is glued to a pair of 2x stringers that raise the table to the level of the sander's platen. The stringers are in turn screwed to the base of the jig. To use all the areas of the belt, we put additional boards on the sanding table to act as spacers. We square the tool to the table with shims of scrap wedged under the sander. A more sophisticated version of this would include a pivoting table that could be tilted into square with the platen.

—Sven Hanson, Albuquerque, N. Mex.

PAINT-GRADE GAP FILLERS

A good way to fill butt joints in molding that will be painted is to cut some long, thin wedges on the table saw. The wedges can then be angled into the gaps, cut off flush with a sharp knife and then glued in place—no need for sloppy caulking or wood filler. Of course it's better to make the joint as thin as a razor line in the first place, but things don't always work out that way.

—David Campbell-Page, Toronto, Ont.

JIGSAW COPING

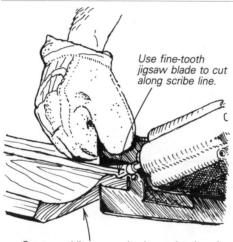

Use fine-tooth jigsaw blade to cut along scribe line.

Crown molding securely clamped to bench

Wide crown moldings are in vogue in our area, and coping a bunch of them by hand gets pretty tedious after a while. To speed things up, we now freehand the cuts with a jigsaw (drawing above). With the workpiece securely clamped to the bench, I use a narrow, fine-tooth, bi-metal blade to follow the cut line. With practice and some caution (I wear a heavy leather glove on the hand guiding the saw's base) good-looking coped joints on wide hardwood moldings are easy.

—Lucian L. Tatum III, Griffin, Ga.

COPING TABLE

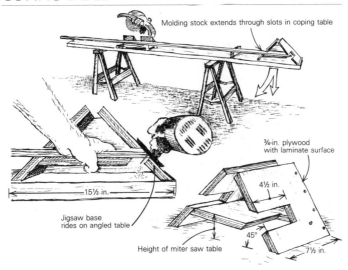

Molding stock extends through slots in coping table

¾-in. plywood with laminate surface

4½ in.

15½ in.

Jigsaw base rides on angled table

Height of miter saw table

45°

7½ in.

As much as I appreciate the delicate craft of coping moldings by hand, I have to agree with Lucian Tatum (facing page) that a jigsaw makes the work go much more smoothly. It can be difficult, however, to hold the foot of the saw in plane with the 45" bevel on the end of the workpiece while following the lines of a complicated, narrow molding profile. The coping table shown in the drawing above solves that problem for me.

The table is a pyramidal box with slots cut into both sides for the molding stock. After beveling a piece of molding on the miter saw, I slide the stock into the coping table. The box's dimensions allow ample hand room for holding the work steady, while my fingers remain clear of the blade. During a cut, the saw rides on the angled side of the coping table. Because I made the table out of the sink cutout from a plastic laminate countertop, the saw glides easily over the plastic surface. I use a Bosch 1581 variable-speed jigsaw with their T119BO coping blade for this work, but any jigsaw with a roller guide and a fine-tooth scroll-cutting blade should work well. For intricate profiles, I use the blade as a power rasp.

—*Grafton H. Cook, Dowagiac, Mich.*

ROUTER-MADE MOLDINGS

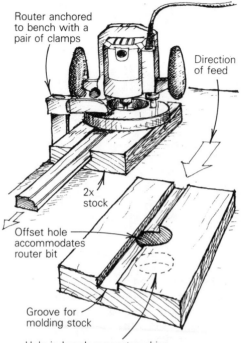

Router anchored to bench with a pair of clamps

Direction of feed

2x stock

Offset hole accommodates router bit

Groove for molding stock

Hole in bench evacuates chips

I needed a special molding to complete a baseboard detail, but my router table was several hundred miles away on another job. Fortunately, the situation forced me to come up with an alternative method for site-milling trim stock. I think my new method is faster, more accurate and safer than using a router table—especially if the moldings are narrow and thin.

As shown above, I used a scrap of 2x stock about 1 ft. long and about the width of my router's base. I cut a lengthwise groove near the middle of the 2x, just a pinch larger than the depth and width of my molding stock. Then I used a hole saw to bore a 1½-in. dia. hole that is offset from the center of the groove. This hole accommodates the router bit, and it should be to the left of the groove as you face the jig. This is to ensure that the router bit, which turns clockwise, will be turning into the work as you feed the stock into it. Next, I bored a similar hole in the top of my jobsite workbench to allow the wood chips an escape route.

I positioned my router over the hole in the jig and anchored the router to the table with a pair of clamps. The clamps were arranged on opposite sides of the router's base, in line with the groove in the

2x stock. By sighting down the groove I could easily adjust the router, both vertically and horizontally, until I had the bit in the exact position that I needed for the molding profile. Cutting the moldings is a simple matter of turning on the router and feeding the stock into the groove. In a few minutes I had hundreds of feet of molding. And because the stock was captured in the groove of the jig under the base of the router, my fingers never got near the cutters.

—Bill Young, Berkeley, Calif.

HOLDING MOLDING

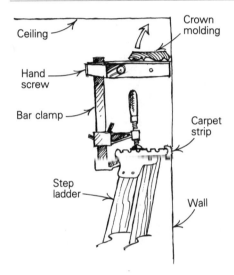

When I had to install some crown molding I began by using a loop of string tacked to the wall near the ceiling to hold up the free end of the molding. The problem was, I kept scratching the freshly painted wall using this method. I had better success with the setup shown here. I used a bar clamp to make a mast that extended beyond the top of a stepladder, and I clamped a handscrew to it to make a horizontal shelf to support the molding. Standing on another ladder, I could easily tilt the workpiece into position and snug its coped end against the adjoining molding while the free end of the molding rested on the handscrew shelf.

—John Ketcham, Rockford, Ill.

STUD FINDING

When trying to find a stud in a wall, get down on your knees and look at the baseboards. In addition to being nailed to the sole plate, baseboards are typically nailed to the studs. By examining the baseboards you can almost always find where the nail holes have been filled.

—John E. Schafer, Columbia, S. C.

NAILING BETWEEN STUDS

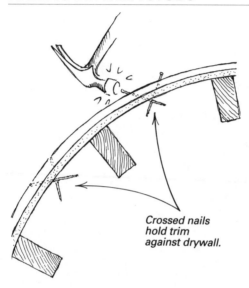

Crossed nails
hold trim
against drywall.

As a general contractor, I occasionally have to install moldings. A recurring problem is the lack of solid backing. Either the stud isn't where I need it, or a curvy wall requires more points of attachment than that provided by the studs. In these situations I affix the trim to the drywall or plaster by driving pairs of nails at opposing angles, as shown above. A nail driven straight into drywall will work itself out, but nails cross-nailed in this manner will hold firmly.

—Simon Lockwood-Menkes, Los Angeles, Calif.

BASEBOARD SHIMS

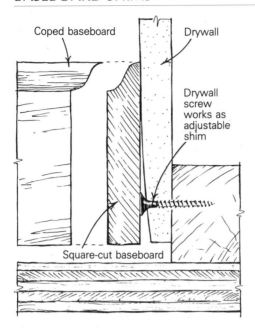

Coped baseboard

Drywall

Drywall screw works as adjustable shim

Square-cut baseboard

It can be frustrating to install precisely fit baseboards over an uneven substrate like drywall. Baseboards often sit atop tapered drywall edges, causing the wood to tilt a bit out of plane with the wall. This can cause an unsightly gap at a corner where a coped baseboard intersects a square-cut baseboard (drawing above).

I avoid this problem by driving $1\frac{3}{8}$-in. drywall screws into the sill plates at each inside corner. The screws only need to be installed under the square-cut pieces of baseboard. As shown in the drawing, the screws work as adjustable shims, allowing me to run them in or back them out as needed to put the baseboard into plane with the wall. I use a short piece of baseboard with a coped end on it to test the corner joints for fit as I install the square-cut pieces. That way I don't have to keep running back to my saw to adjust the coped cut. For outside corners, I put a screw on each wall.

—Ralph W. Brome, Greensboro, Md.

COPING WITH ROUNDED TRIM

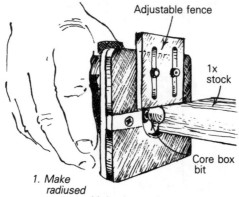

Adjustable fence

1x stock

Core box bit

1. Make radiused groove with laminate trimmer.

2. Remove back edge with the table saw.

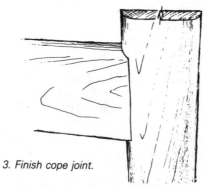

3. Finish cope joint.

Our client's house has some arched windows and doorways, and to remain in keeping with these openings the trim pieces have edges that have been radiused with a ⅜-in. roundover bit. The rounded trim adds a nice touch, but when it came time to join chair rails and baseboards with the door trim, we found ourselves with a problem to solve. My partner, Paul Hannenmann, came up with the solution above.

First he made an adjustable fence out of a piece of ¼-in. acrylic for his Porter-Cable laminate trimmer. The fence has slots in it that allow it to be moved in relation to the bit. The screws provided on the tool's base hold the fence in place. Using a ½-in. core box bit, Paul plowed a radiused groove in the end of a piece of 1x pine trim (drawing 1). He adjusted the fence to leave a paper-thin edge on the exposed side of the trim. Then he used a table saw to remove the back edge of the groove (drawing 2). The finished reverse curve allows the base or chair-rail member to abut the door casing without a gap (drawing 3).

—Anthony Patillo, Conway, Mass.

SLOTTING BASEBOARDS

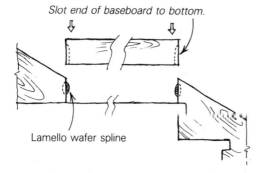

Slot end of baseboard to bottom.

Lamello wafer spline

I recently had to install a piece of a 1x4 baseboard on a landing between two flights of stairs. Although both skirtboards were already installed, I wanted to use my Lamello joining tool to ensure a flush fit at the butt joints.

As shown in the drawing above, my solution was to install the Lamello wafers in the skirtboards in the normal fashion. Then I cut corresponding slots in the ends of the baseboard, extending them to the bottom edge. This allowed me to slip the wafers into the skirtboards, and then slide the baseboard downward for a tight, smooth fit.

—Will Milne, San Francisco, Calif.

MATCHING MOLDINGS

When remodeling, sometimes you need a short piece of trim that is no longer available at the lumberyard. Take a look in the closets—you might find just enough to suit your needs.

—*Craig Horstmeier, Davis, Ill.*

REDUCING SPRINGBACK

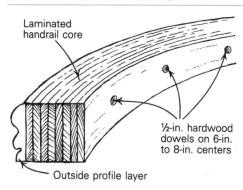

Laminated handrail core

½-in. hardwood dowels on 6-in. to 8-in. centers

Outside profile layer

I made some curved handrails by laminating a number of thin strips of wood together using aliphatic resin glue. They came off the mold with the right radius, but after lying around the job site for a few days, they started to lose their curve as the laminations began to creep along the glue lines.

As shown in the drawing above, I virtually eliminated the problem in subsequent rails by inserting ½-in. hardwood dowels (liberally coated with glue) into the core of the rail on 6-in. to 8-in. centers. The dowels are hidden by the outer, profiled laminations. This method also eliminates the danger of cutting into metal screws when installing balusters or cutting the handrail to fit posts, goosenecks, turnout easings and other fittings.

—*Klaus Matthies, Regina, Sask.*

HANDRAIL JOINT

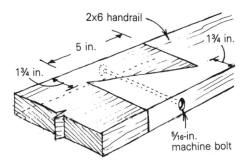

2x6 handrail

5 in.

1¾ in.

1¾ in.

⁵⁄₁₆-in.
machine bolt

To make the handrail joint shown above, I start by laying out and cutting the first piece, then using it as a pattern for the second piece. Next I temporarily clamp the two together and improve the fit by running a saber saw with a fine-toothed blade through the tight parts of the joint until I'm satisfied with the fit. For 2x6s, I use a ⁵⁄₁₆-in. by 5-in. machine bolt (countersunk on both edges) to hold the pieces together. If I'm making more than one joint, I first make a pattern out of aluminum flashing or cardboard.

—Les Watts, Herndon, Va.

SHELF SUPPORT

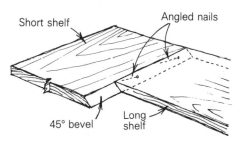

Short shelf

Angled nails

45° bevel

Long shelf

I was recently asked to add an upper closet shelf to each of two homes a few blocks apart. Both houses were built in the 1930s. The first had an existing U-shaped shelf in a 42-in. wide closet. The short return sections were secured to the longer one with a 1x4 cleat fastened under the joint. This seemed like a decent solution to the problem, so I did the same with the new upper shelf.

The next day I went to the second job site. There I found a clever solution to the same situation. The short shelves had been secured first, and the lead edge of each one had been beveled at a 45°

angle with the point of the bevel on the bottom of the shelf. As shown in the drawing on p. 123, the ends of the long shelf were beveled in the opposite direction, allowing it to rest snugly on the shorter shelves. A little glue and some 4d finish nails driven at an angle completed this simple, strong joint.

—David Strawderman, Los Angeles, Calif.

CUTTING CLAPBOARDS

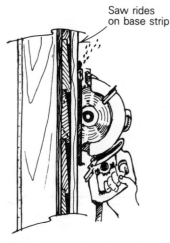

Saw rides
on base strip

When remodeling or building additions, it is often necessary to cut existing clapboard siding along a vertical line. Problem is, the saw's shoe wants to hang up on the protruding lower edges of the clapboards, which in turn makes the blade dip into the sheathing underneath it. To ensure a cut of even depth, I tack a ¾-in. thick base strip to the siding, as shown above. Then I set the saw to cut a depth equal to the thickness of the base strip plus that of the siding. As the saw's foot rides on the base strip during the cut, a constant depth is maintained. To make the cut even more accurate, you can tack a guide for the side of the saw's shoe to the base strip.

—Loran Smith, Dover, N. H.

SHINGLE TLC

I replaced some windows in a 30-year-old house that was covered with cedar shingles, and try as I might I found it impossible not to scar some of the shingles with my ladder. Also, I had to remove some of the shingles to trim them, which further damaged their faces.

After thinking about the problem, I decided to try reviving the shingles with a very soft wire brush. I brushed with the grain, starting at the top and working down. I first tried a small spot in an unobtrusive place, and found that the shingles had a nap like suede or velvet. Brushing with the grain pushed all the fibers back into the grain lines that emerge as the shingles weather with age. I brushed all the shingles around each window about a foot out from the casings, and then I sprayed them lightly with a hose to clean the wall of fine particles. When everything dried, it was barely noticeable that any work had been done.

—*Richard E. Reed, Doylestown, Pa.*

DRYING RACK FOR SIDING

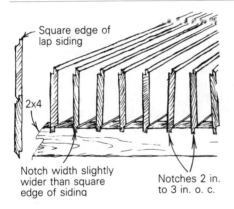

Square edge of lap siding

2x4

Notch width slightly wider than square edge of siding

Notches 2 in. to 3 in. o. c.

By the time we got the sheathing on our house it was winter—too late in our rainy part of the world to take a chance on putting unfinished siding on the walls. So to keep the work progressing, I built a number of drying racks out of scrap 2x4s. As shown in the drawing above, they have notches (cut with a dado blade) at 2-in. to 3-in. intervals. The notches accept the squared top edge of the lap siding. I arranged these racks in tiers on about 6-ft. centers in our unfinished house. This enabled us to prime both sides and paint one face of the siding while under a finished roof.

—*Jim Gurman, Arcata, Calif.*

SIDING HOLDUP

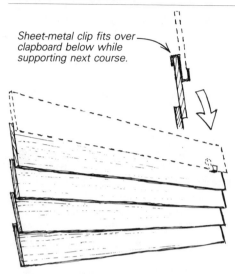

Sheet-metal clip fits over clapboard below while supporting next course.

One night I went to bed dreading the next day. Working solo, I'd be siding the gable end of a house with clapboards, my only helper on the ground cutting and handing me the material. Like magic, the solution to my problem came to me in a dream that night.

As shown in the drawing above, I made an S-shaped clip out of a 4-in. sq. piece of sheet metal. In use, the top leg of the clip hangs over the prior course of siding, while the bottom leg supports the next course. The arrangement allowed me to get a few nails into each course to keep it aligned before pulling the clip out and moving it up. My "Trip Clip" also proved to be handy for supporting the end of my tape measure over long runs.

—Trip Renn, Chapel Hill, N. C.

SIDING HOOKS

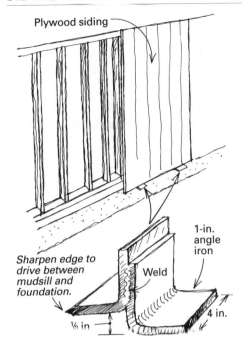

Plywood siding

1-in. angle iron

Sharpen edge to drive between mudsill and foundation.

Weld

4 in.

½ in

When putting up plywood sheathing or siding, I use the angle-iron hooks shown here to support the bottom edge. They keep the bottoms straight, and the ½-in. offset ensures that the plywood will overlap the foundation. To use the hooks, I drive them between the mudsill and the foundation, place the panel and nail it along the edges for position. Then I pry the hooks out with my hammer's claw. I painted them yellow to keep from losing them in the mud.

—*Roy Rider, Corvallis, Ore.*

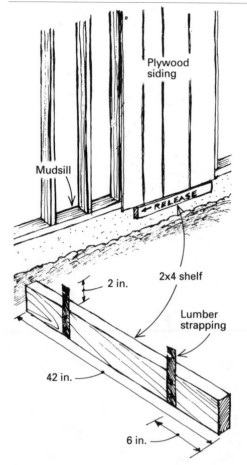

When we built our house we dreaded the thought of installing the 70 sheets of 4x9 T 1-11 siding. Then we saw Roy Rider's tip on siding hooks (p. 127), which combines a couple of pieces of welded angle iron wedged under the mudsill to support the siding as it's nailed to the framing. Lacking a torch, we modified Rider's idea and built the shelf shown above.

The body of the shelf is a 42-in. 2x4. The two metal straps affixed to it were made from the steel banding that wrapped our framing lumber. We drilled each strap with three small holes, then with tin snips modified the top hole of each strap into a slot. The straps extend 2 in. above the 2x4. Each strap is affixed to the mudsill with a short roofing nail.

To use the shelf, we tacked it to the mudsill through the slot and perched a sheet of siding on it. Then we attached the siding to the framing with a couple of nails at the top of the sheet. With the siding in the right place, we knocked the shelf off its nails (we marked it with a bold "release" sign to remind us which way to hit it) and bent the siding outward a bit to allow access to the nails so that we could lever them out with a pry bar.

—*Jim Finnegan, Tazewell, Tenn.*

HANGING FASCIA BOARDS

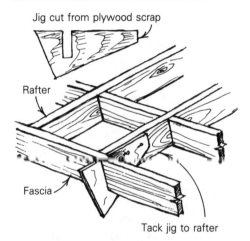

Jig cut from plywood scrap

Rafter

Fascia

Tack jig to rafter

It usually takes two people to hang fascia boards. Even then it can be pretty precarious out there on the end of a rafter, straining to support a heavy fascia board with one hand, while trying to line up a mitered corner and sink a galvanized 16d nail with the other hand. With the help of the simple jig in the drawing above, even one person can do it.

I tack one jig near each end of the fascia. I drive the nails just far enough to support the weight of the fascia. Then I lower the fascia into the slots in the jigs (the slots should be a little oversize to prevent binding). The jigs hold the fascia in approximately the right place while I adjust it for alignment and nail it in place.

—*Neal Bahrman, Ventura, Calif.*

MUDBUCKET BOOTS

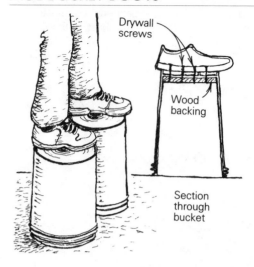

Drywall screws

Wood backing

Section through bucket

A carpenter friend of mine doesn't have any trouble reaching a typical ceiling when he needs to install insulation, tape drywall or nail on the crown moldings. As shown in the drawing above, he uses drywall screws to affix a pair of old work shoes to the bottoms of a couple of empty joint-compound buckets. The shoes are screwed around the perimeters of the soles, through the bottom of the bucket and into a piece of wood backing.

The bucket boots add about 15 in. to your height, and they are surprisingly stable. One drawback is that you must step through narrow doorways one foot at a time when wearing the big feet.

—Stephen M. Kennedy, Orrtanna, Pa.

8

PAINT, CAULK AND GLUE

STEADY THAT BUCKET

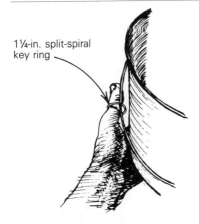

1¼-in. split-spiral key ring

Attaching a 1¼-in. split-spiral key ring to the bail of your paint pot makes holding a bucket all day a lot easier. Support the bottom of the bucket with your hand, as shown above and slip your thumb through the ring to steady it.

—*Bryan Humphrey, Wilmington, N. C.*

PAINTED-TRIM PREP

Sometimes when a house interior needs repainting, the woodwork is too rough to take a new coat of paint, but not bad enough to warrant stripping. Under these circumstances, I prepare the woodwork by using a sponge, a bucket of water, a scrub brush and some wet/dry sandpaper. I keep the sandpaper and work surface wet throughout the process, and I use my hands or the sponge as a sanding pad. I find that the paper will last up to an hour if I periodically use the brush to scrub the paper in the bucket.

Using this wet process keeps the dust suspended in the water rather than in the air. The resulting surface is quite smooth to the touch, but it still has the necessary roughness for good paint adhesion. I've used this technique with old oil-base and water-base paints with equally good results.

—John Glenn, Brookline, Mass.

PAINT SPOUT

Though they are available for gallon cans, I have never been able to find a pouring spout or a flexible plastic lip for quart paint cans. Therefore, when I pour paint from a quart can I always have a terrible time cleaning the groove at the top of the can that mates with the lid. To eliminate this problem, I twist a paper towel into a rope that will fit tightly into the groove. Then I pour my paint over the twisted towel, which acts as a makeshift lip and remove the towel. A quick wipe with another piece of paper towel leaves the groove clean and ready for its lid to be tapped back in place.

—Kenneth G. Warr, Trussville, Ala.

PAINTING TRAYS

The lids of 5-gal. joint-compound buckets make great trays for 1-gal. cans of paint. In addition to serving as a platform for the paint can, the lid is large enough to hold a paintbrush, and the lid's lip is high enough to prevent drippings from overrunning the tray.

—Steven Silkunas, Philadelphia, Pa.

WIPING PAINT BRUSHES

Bent wire conforms to rim.

Painters often take the excess paint off their brushes by wiping the sides of the bristles on the inside lip of the can. Eventually the channel around the rim fills up, and paint runs down the outside of the can. Some painters avoid this problem by poking holes in the bottom of the channel with a 16d nail, the theory being that the paint will drip back into the can before filling the channel to overflowing. My solution to this problem avoids the rim altogether.

As shown in the drawing above, I bend a stiff piece of wire (such as a coat hanger) into a crossbar that spans the can. The wire hangs on the rim of the can a little off center to make the paint easy to reach with my brush. I wipe the excess paint on the wire, and the paint simply drips back into the can without touching the rim. The crossbar is easily removed for cleaning and reuse. If I'm using a paper or plastic container, I poke holes in the sides near the top and run a straight wire through them.

—Michael R. Hogan, St. Croix, V. I.

PAINT-TRAY LINERS

The big plastic bags that are being used nowadays by grocery and discount stores fit nicely over a 9-in. paint tray, making a durable liner that keeps cleanup time to a minimum. You do, however, need to be sure to turn a printed bag inside out because the ink can contaminate your paint. In my experience, the bag liners work well with latex and oil-base paints.

—Kenneth G. Warr, Trussville, Ala.

PERPETUAL PAINTBRUSHES

When using oil paints, I don't enjoy cleaning the brush at the end of the day, only to get it dirty again the next morning. So between uses, I simply stick the brush in a can of water. The water keeps the air away, and thus keeps the brush from setting up. This works for varnish and stain as well. When I'm ready to use the brush, I just give it a few heavy strokes against a clean cardboard box. Oil and water don't normally mix, so what little water remains on the bristles is rapidly absorbed into the cardboard.

At the woodshop where I teach, I keep a perpetual brush dedicated to varnishing projects in a 1-lb. coffee can full of varnish. The lid has an "X" cut in it, through which the handle protrudes. This brush is used daily in varnishing projects, and the varnish is replenished on a daily or weekly basis. As long as the brush is used at least once a month, there have been no problems.

—Mark White, Kodiak, Alaska

COSMETIC ADVICE

One of my first jobs in the trades was on a painting crew that specialized in painting Victorian houses. Two of the techniques that I learned then have stayed with me through years of carpentry and contracting, and I'm always surprised that more contractors don't know them. The first is to caulk baseboards. On all paint-grade work, a bead of caulk at the wall and floor junctures is miraculous at cleaning up the lines of an entire room. A neat bead of caulk at an existing oak floor makes the baseboard and floor look almost new.

The second technique is to use the long, narrow paint roller know variously as "slim Jim" or "long John." The handle is 2 ft. long, and the roller is 1 in. in diameter by 7 in. wide. With one of these you can reach behind toilets and appliances and roll right into corners without having to "cut in" with a brush. There is a trick though—the small rollers are notorious for leaving lint behind. This is cured by swiftly passing them over a gas-stove burner or butane lighter before using them. The flame singes the loose lint off the roller—but be careful not to ignite the thing.

—Kurt Lavenson, Berkeley, Calif.

PAINT CADDY

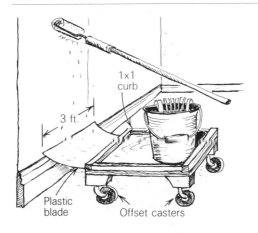

I paint houses for a living, and there are three things about painting with a roller that really annoy me. One is masking the baseboard and spreading out a tarp to catch drips and speckles. Another is moving the paint bucket from station to station, and the third is moving the tarp again, which inevitably results in drops of wet paint smearing the carpet or floor.

Instead of this sequence, I now use a plywood platform on wheels both to carry the paint bucket and to protect the baseboard and floor from paint splatters. A plastic blade on one end of the platform butts up against the wall to catch drips, as shown in the drawing above. To make the blade, I used a section cut out of an old plastic garbage can, and I attached it to the caddy so the curve points upward. This directs paint away from the wall and accommodates the higher baseboards.

—Stan Lucas, Redmond, Wash.

FOILING PAINT DRIPS

To protect items such as hinges, doorknobs or wall phones when I'm painting, I simply wrap them with aluminum foil. It can be molded quickly to fit even the oddest shapes, and it stays put while I wield the brush.

—Marie Christopher, Brady, Tex.

TIGHT-SPOT PAINTING

An inside corner tool used for drywall finishing can also be used as a painting shield around carpeted stairs. The tool fits neatly into the corner of the step (where the riser meets the tread). If the stair is carpeted, the tool is sturdy enough to compress the carpet.

—Kenneth G. Warr, Trussville, Ala.

REMOVE PAINT FROM SCREWS

When a screw slot is clogged with paint, it's very difficult to get a bite on it with your screwdriver. When I'm up against this problem, I chuck a 4d finish nail in my electric drill and excavate the caked-on paint with the point of the nail. When the nail dulls, which doesn't take long, I touch up the point with a few licks from a file. This method works equally well on Phillips-head or straight-slot screws.

—T. D. Culver, Cleveland Heights, Ohio

MIXMASTER

If you are about a week away from using a can of paint that's been sitting around long enough for the solids to settle, put it in the back of your car or truck. Make sure the lid is on tight, turn the can upside down, and let the motion of the road mix the paint as you drive.

—Eric Olsen, El Cerrito, Calif.

PAINT STRAINER

Window screen

Bucket lid with center removed snaps over bucket rim.

Cut away assembled strainer.

Plastic paint bucket

As a painting contractor, I use a lot of paint in 5-gal. buckets. Bits of joint compound and debris from the job site inevitably find their way into the paint, so I strain my paints each day before applying them. I could buy strainers, but I prefer to make my own (as shown in the drawing above).

Using an old plastic paint bucket with a lid, I lay some window screen across the open bucket top and push it downward with my fist to form the strainer. Next I cut a big hole in the lid, leaving just the lock ring. Now I snap on the lock ring, and cut the assembled strainer away from the bucket. Presto! I've got a strainer that perfectly fits 5-gal. buckets.

—Bob Simpson, Cottage City, Md.

HELP FOR SPRAY-CAN NOZZLES

I wanted to use a partially consumed can of spray paint the other day, but the perennial problem prevented me from doing so—the nozzle was full of dried-up paint. I searched around the shop for another nozzle and ended up using the one off my can of WD-40. When I put the nozzle back on the can of WD-40, I gave it a squirt to clean it out, and realized that I can do the same thing with my spray-can nozzles. Now when I'm done using a can of spray paint, I put its nozzle on the WD-40 can for a cleansing blast.

—Charles Buell, Washington, Maine

ADVICE FOR FENCE PAINTERS

A three-story stone home in my neighborhood has stood in stately elegance for 80 years, surrounded by a 4-ft. high wrought-iron fence atop a 2-ft. stone wall. The iron pickets are about 6 in. apart, and they have always been painted black. Recently, the house changed hands and the new owner painted the fence white. The house virtually disappeared.

I think this illustrates a frequently ignored design principle. The impact of a fence or a latticework screen can be heightened or reduced by its color. If you want to see the fence, paint it a light color. If you want to see what is behind the fence, finish it with a dark stain or paint. If you want to eliminate what's behind the fence, paint the fence white. This may seem obvious, but when you start looking at fences it becomes apparent that this is a poorly understood rule.

—Frederic E. Bishop, Farrell, Pa.

STAIN CONTROL

If you've got a wood-staining job coming up, consider using a pad painter to apply the stain with. The foam pad holds more stain than a brush, and the flow can be controlled by the amount of pressure that you apply. I have found that with a little practice, I can almost eliminate the need for wiping the stain with a rag.

I use an old plastic dishwashing-soap bottle to deliver stain to the pad. It has a pop-up cap so it's easy to close, thereby eliminating the chance of a major spill. To keep the stain "stirred," I shake the bottle now and then.

—Emily Neal, Alfred Station, N. Y.

DRIP-CATCHING CUFFS

In a kinder world, we would always be able to apply wood finishes to horizontal surfaces at bench height. But in real life we invariably find ourselves reaching over our heads with a brushful of stain or oil. So you know as well as I the misery of cold, sticky fluid dribbling down over your wrist. Next time try this trick.

Get a rubber glove, of the type sold for washing dishes, for your painting hand. Turn up a 2-in. to 3-in. cuff on the glove, and stuff it with toilet paper—you want a puffy doughnut of tissue filling the cuff and circling your wrist. The tissue holds the cuff out to catch the dribbles and absorbs the fluid so it won't leak out when you lower your arm.

Finally, wrap the end of the brush handle with tissue and secure it with a rubber band. You are now ready to apply the finish. When the tissue rings become saturated, squeeze them out over the finish container to put the liquid back where it can be used again.

—*Jerry Azevedo, Friday Harbor, Wash.*

DRYING RACK

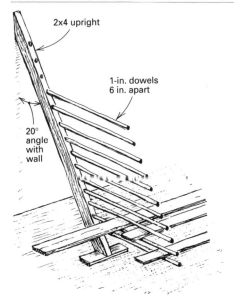

2x4 upright

1-in. dowels
6 in. apart

20°
angle
with
wall

We preprime and varnish a lot of the trim we install in houses, and finding a good place to let the material dry without getting in the way used to be difficult. To solve the problem I came up with the drying rack shown here. I make the brackets out of standard 2x dimension lumber. Using a drill press, I drill 1-in. dia. holes 6 in. apart for 3-ft. long, 1-in. dowels. As shown in the drawing, the brackets lean against the wall at a 20° angle, and the holes are oriented at a 70° angle so that the dowels end up level while the racks are in use.

I space the racks about 6 ft. apart and start loading them from the bottom up. It's amazing how much material they will hold and how strong they are. I've even used them to hold freshly painted doors.

—*Daniel E. Perry, Vineyard Haven, Mass.*

EXTENDING VARNISH SHELF LIFE

Air trapped inside a can of wood stain or varnish causes the partially used portion to skin over and eventually solidify. My solution to this wasteful problem is to displace the trapped air with clean, smooth hard pebbles—the kind found in creekbeds. I just drop them into the can until the liquid rises to the can's rim, then replace the lid. I've got ready-to-use varnish or stain the next time I need it.

—Betsy Race, Euclid, Ohio

CAULK SAVER

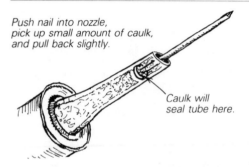

Push nail into nozzle, pick up small amount of caulk, and pull back slightly.

Caulk will seal tube here.

It would be nice if all caulking jobs required exactly the amount of caulk that comes in a tube, but they don't. Instead, we are left with partially consumed tubes of caulk that are worth saving. Some people put a nail in the end of the tube, and put it away for future use. The next time they need the caulk, they pull out the nail and try to squeeze the caulk out of an ⅛-in. dia. nail hole. Unless you've got forearms like Popeye, this is difficult.

I also use a nail to plug a caulk tube, but I insert it head first, as shown in the drawing above. I swirl the nailhead around in the caulk and then I pull the nail back out until it seats against the inside of the nozzle. The caulk on the shank of the nail will make a seal that protects the contents of the tube. When I pull out the nail, the caulk seal comes with it, and the goo inside the tube is still usable.

—Norm Jespersen, Royersford, Pa.

CAULK-TUBE CAP

Instead of plugging a partially used tube of caulk with a nail or capping it with a wire nut, I keep the remaining contents from hardening with a piece of Romex wire sheathing. First I pull the wire out of a 5-in. or 6-in. Romex offcut. Then I shove the plastic sheath over the caulk-tube nozzle and pump a little caulk into the sheath. When I pull it off the nozzle weeks later, the rest of the caulk in the tube is still fresh. The sheathing also works as a nozzle extension for getting at hard-to-reach crevices.

—G. Morrison, Spring Mills, Pa.

NOZZLE DECLOGGING

In my experience, it doesn't matter if I insert a nail or a screw into the nozzle of a partially used tube of caulk. The caulk will still coagulate into a frustrating lump that renders the tube useless. To remove the hardened caulk, I use my utility knife to slit the entire length of the nozzle. Then I pry out the caulk lump and tape the nozzle back together with electrician's tape.

—Nancy Hart Servin, Oakland, Calif.

CAULK CORK

Caulk left to harden

I used to have a problem with my caulking gun being clogged every time I wanted to use it. Now, when I'm ready to put it away, I pump out a short rope of caulk and leave it there, as shown above. It hardens into a plug that I can pull out when I want to use the remaining caulk.

—Howard Moody, Upper Jay, N. Y.

WARMING CAULK TUBES

When the weather turns icy up here in New Hampshire, construction adhesives and caulks get so cold they won't budge out of their tubes. To prevent this, I installed a metal can (about one gallon in size) under the hood of my truck. I placed it near, but not quite touching, the engine. The lidless can has enough room in it for five tubes of caulk or adhesive. On my drive to work in the morning, the tubes warm up enough to allow their contents to flow easily. Depending on how cold it is, the heat from the mass of the engine block keeps them warm for several hours.

—Brian Carter, Concord, N. H.

DRIP-FREE CAULKING

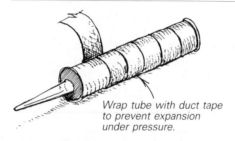

Wrap tube with duct tape to prevent expansion under pressure.

After years of frustration trying to apply a neat line of caulk, I figured out how to make the caulk stop discharging the instant I stopped squeezing the trigger. The secret is to wrap the tube with duct tape, as shown in the drawing above. This prevents the tube from expanding under pressure and then collapsing (and discharging caulk) when the trigger is released.

Besides permitting precise control of the caulking job, I save on caulk. If I'm careful, it's possible to work without a cleanup rag.

—Keith Ojala, Birmingham, Mich.

SMOOTH CAULK JOINTS

Leaving a smooth, attractive line of caulk is a difficult task for me—even when I follow the manufacturer's recommended procedures. I've used rounded pieces of wood and plastic, and my finger, all both wet and dry. The result is invariably a caulk seam of amateur status.

For me, the solution is an ice cube. I take an ordinary ice cube and shape it to a rounded tip using the palm of my hand. Then I wipe down the entire length of the caulk seam with the cube. This leaves a perfectly smooth seam, and the excess caulk is easily removed from the cube. Water isn't the best tooling liquid for all caulks, though, so check the labels.

—*William A. Rolke, Columbus, Ind.*

SUBSTITUTE CAULK GUN

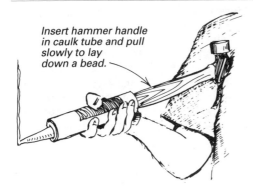

Insert hammer handle in caulk tube and pull slowly to lay down a bead.

I had my doubts about whether or not this technique was going to work, but after climbing three icy ladders to get to a skylight that needed caulking, I was willing to give it a try. You see, I'd remembered the caulk but left the caulk gun in the truck. And I sure didn't want to hike back down those slippery ladders unless I had to.

So I cut the end of the tube's nozzle, punctured the membrane and put the long handle of my framing hammer (drawing above) into the plunger end of the tube. I placed the hammer head in my armpit, and by simply pulling on the tube, I laid down a very respectable bead of caulk. In fact, I found the control and the accuracy of this method to be nearly the equal of my caulk gun.

—*Robert Hausslein, Andover, Vt.*

REMOVING CONTACT CEMENT

If you've been working with contact cement and got some on your hands, use a crepe rubber sanding belt cleaner to take it off. It works like a big eraser to remove the dried glue with just a few strokes.

—*Diane Hartley, Klamath Falls, Ore.*

LAMINATE SPACERS

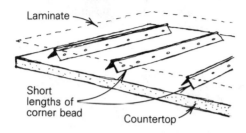

Using contact cement is a tricky business. When using it to glue down laminates to countertops, I use short lengths of drywall corner bead to separate the two glue-coated surfaces until I'm sure of correct placement (drawing above). Then, by removing them one at a time, the laminate can be carefully and accurately pressed into place. The corner-bead scraps are easy to keep clean and can be stacked in a compact pile for easy transport and storage.

—Larry Wilson, Uniondale, Pa.

PLASTIC LAMINATE ON TWO SIDES

To apply plastic laminate on both faces of a cabinet door, you can save yourself some time by doing them simultaneously. Drill a hole at each corner of the door, 1 in. to 2 in. from the edges, and drive a small wood screw a few turns into each hole. Apply contact cement to the side with screws, then turn it over so that the door is supported on the screw feet. Apply cement to the other side of the door, and to the laminated pieces. When the cement is tacky, apply the laminate to the side without the screws. Flip the door, take out the screws, laminate the second side and you're ready to finish the edges with a trim bit in a router.

—Roy T. Higa, Honolulu, Hawaii

9

PLUMBING, BATH AND ELECTRICAL

BALLOON PLUG

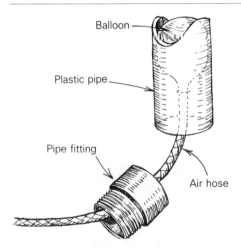

Balloon

Plastic pipe

Pipe fitting

Air hose

I had to glue a threaded fitting onto a piece of plastic pipe that kept dribbling water onto the bonding surfaces. To slow the flow, I got out the bike pump and an ordinary balloon. As shown here, I slipped the balloon over the end of the air hose and ran the hose through the fitting and into the pipe. Inflated, the balloon stopped the dribble long enough for me to dry off the parts and glue them together.

—Ed Self, Los Gatos, Calif.

ICE VALVE

When you have to cut a water line in a situation where it's impractical to shut off the water, surround the pipe on each side of the cut with dry ice. The water adjacent to the ice will freeze. Make the cut, insert the fitting and thaw the pipe.

—Jack H. Gillow, Highland, Mich.

FLUX-BRUSH CONTAINER

As all good plumbers know, one of the secrets of a successfully sweated joint is cleanliness. And one of the hardest things to keep clean is a flux brush. If you keep it in a toolbox, it just collects dirt.

I keep my brushes clean in storage tubes made of ¾-in. PVC pipe. For each brush, I cut a piece about 7 in. long, and close its ends with a couple of PVC caps. Now my brushes stay clean, and the white containers make them easy to find.

—Alan Mendelsohn, Indianapolis, Ind.

PLASTIC PIPE CUTOFF JIG

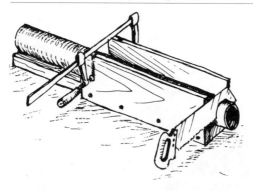

On a recent plumbing job I had to cut the ABS pipe with great accuracy. Skewed cuts would have thrown off my layout, making it tough to fit the parts into the tight existing conditions. As shown above, I used two 1x4s and a 2x4 to make a cutoff jig. The width of a 2x4 handily accommodates a 3-in. dia. ABS pipe. And by adding spacers, I can use the jig for 1½-in. and 2-in. pipe. A bar clamp at one end pinches the fences inward, keeping the pipe from spinning as it is cut.

—Duff Bogen, Seattle, Wash.

REMOVING BROKEN PLASTIC PIPES

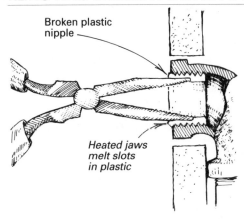

Broken plastic nipple

Heated jaws melt slots in plastic

I had to remove a threaded plastic nipple that had broken off inside an elbow where it came out of the shower wall. To get the nipple out, I heated the jaws of a pair of needlenose pliers with a propane torch and poked them into the broken nipple (drawing above). Then I slowly opened the jaws, melting slots in the plastic for the pliers to bear against. When the pliers cooled down and the plastic solidified, I spun the pliers with the jaws open to remove the broken part.

—*Sandy Tod, Campbellville, Ont.*

FINDING A PUNCTURED PIPE

On a couple of occasions during my career as a home remodeler I have accidentally driven nails or drywall screws into copper plumbing lines. Sometimes the leak will begin to seep days, weeks or even months later as electrolysis causes the hole to enlarge or the fastener to corrode. I've found these situations to occur most often while nailing base trim on walls that hide plumbing for baseboard heaters. Damp spots on the floor, wall or downstairs ceiling may be the only indication of trouble. The problem can be fairly easy to fix; the hardest part is finding out which nail did the damage.

To solve this problem, I use a long set of probe cables for my volt/ohm meter. I test for continuity by plugging one lead into the ground terminal of a nearby outlet. I then use the other probe to test the head of each fastener until I find one that completes the circuit. Remember that it may be necessary to wait until the area dries out somewhat to be sure that the wood or drywall itself isn't wet enough to conduct electricity on its own.

—*Chris Dowd, Biddeford, Maine*

APPLIANCE DRAINS

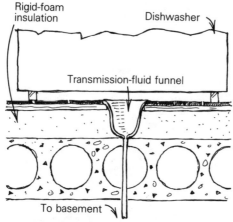

Rigid-foam insulation

Dishwasher

Transmission-fluid funnel

To basement

Most of the houses I build have precast-concrete plank floors covered with rigid insulation topped with plywood. The plywood serves as a subfloor for carpet, hardwood or tile. As you can imagine, water leaking onto such a floor could cause some real damage: if it made its way under the foam, it would take forever to evaporate.

Dishwashers are notorious for leaking around their seals, or if the door doesn't shut tightly. So under the dishwasher in a house I worked in, I inset a plastic funnel, as shown above. The funnel is the type sold at auto-supply stores for transmission fluid. I used a

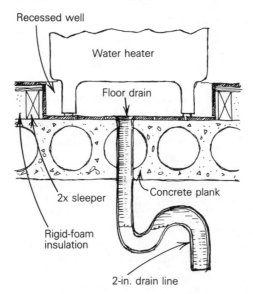

Recessed well

Water heater

Floor drain

2x sleeper

Concrete plank

Rigid-foam insulation

2-in. drain line

hole saw and a rasp to excavate a cavity in the foam for the funnel and sealed around its lip with silicone caulk. Then I drilled through the concrete plank floor for a drain tube that daylights near a floor drain in the basement. I chose not to run a sealed line so that any leakage could be easily spotted.

Plastic pans that are designed to go under hot water heaters can be used, but they are shallow and I have my doubts about their usefulness if a tank should fail. I installed the drain I made under the water heater by stopping the foam insulation shy of the water-heater tank (bottom drawing, facing page). I lined the resulting well with tile and placed a 2-in. drain line in its center. To prevent air infiltration, I put a trap on the line below the concrete plank. This drain also daylights in the basement where any runoff can be spotted immediately.

—Joseph Fetchko, Ocean City, Md.

FIBERGLASS TUB FIX

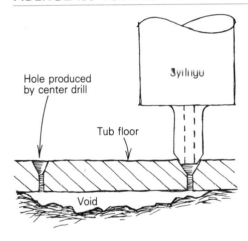

Hole produced
by center drill

Syringe

Tub floor

Void

Galveston Island endures strong winds and hurricanes, so the houses are built on pilings; when the wind blows they flex a bit. Reasoning that this flexing would crack any tile work that I put on the walls of my house, I thought I was smart to install one-piece fiberglass tub/shower enclosures in my two bathrooms.

I like a shower or tub floor to feel solid, so before setting the tubs, I placed 10 gallons of mortar under each one for complete bearing. However, the enclosures had voids between the floor of the bathtub and the chopped fiberglass base. I could feel the voids flex as I stood in the tub, and soon the floors began to crack. Rather than replace the tubs, I came up with another solution.

First I cleaned and waxed the tub floors with car wax. Next I used a center drill to bore holes at each end of each void in the tub floors, as shown in the drawing on p. 149. Center drills combine a countersink and a drill bit and are available for a few dollars at any machine tool supplier. Then I used a syringe with the needle removed to inject fiberglass resin into one hole until it emerged from the adjacent hole. After wiping up the excess, I put a couple of drops of white-pigmented resin from my local boat supply atop the holes. Then to avoid any sanding, I covered each hole with a postage stamp-size piece of wax paper. After the resin hardened, I pulled away the wax paper. The excess resin flaked away easily because the car wax prevented it from bonding to the tub bottoms, and the wax paper left behind a smooth surface. The fix took just an hour and a half.

—*Will Rainey, Galveston, Tex.*

SUPPORTING SINK CUTOUTS

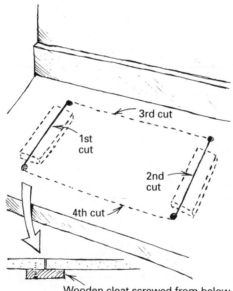

3rd cut

1st
cut

2nd
cut

4th cut

Wooden cleat screwed from below

When cutting the hole for a sink in a laminated countertop, care must be taken to keep the cutout from falling. If it falls, the cutout can tear away a patch of the laminate. The drawing above shows the method I use to support the cutout when I'm working solo.

First I drill holes at the four corners to accommodate my jigsaw's blade. Then I make cuts #1 and #2 on both sides of the cutout. I screw a pair of cleats from below to the underside of the countertop to support the cutout at the ends. When I've made cuts #3 and #4, I remove the cutout, square its edges and trim it for use as a portable counter.

—*Peter Campaner, Thunder Bay, Ont.*

SHOEHORNING A VANITY TOP

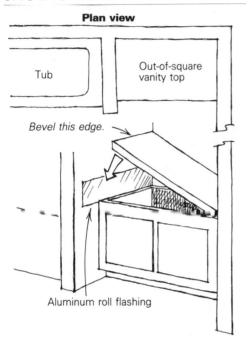

Plan view

Tub

Out-of-square vanity top

Bevel this edge.

Aluminum roll flashing

Every so often I have to install a vanity top between two out-of-square walls. If the opening is wider in front, it's not a problem because I can slide the top in easily. But if the backside is wider than the opening, it can be tough to install the top and maintain a tight fit—there should be no more than a $\frac{1}{8}$-in. gap between the top and the wall.

In a case like this, I bevel the countertop along the bottom edge of one side, as shown above. Next I lean a strip of aluminum roll-flashing against the wall on the side where the beveled edge will rest. The flashing protects the drywall as I place the top on its base, using a little persuasion. To finish up, I pull out the flashing and caulk the edges.

—*Rick Morgan, Manchester, N. H.*

ELECTRICIAN'S STICK

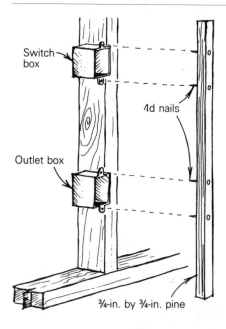

Switch box

4d nails

Outlet box

¾-in. by ¾-in. pine

The next time you have to affix electrical boxes to stud framing for switches or outlets, try using the jig shown above to position the boxes. I make the jig out of a strip of ¾-in. by ¾-in. pine. Two pairs of 4d nails driven through the stick correspond to the threaded holes in the boxes. To use the stick, position the boxes over the nails and place the stick on the floor adjacent to the stud. Now you can nail the box to the stud, and the height will be right every time.

—*Santo A. Inserra, Jamestown, N. Y.*

GLUING OUTLET BOXES

My partner and I recently remodeled an old home. Once all the new switch and outlet boxes were cut into the lath-and-plaster walls, we noticed some were not as secure as we would have liked. So we slipped the straw-shaped nozzle on a can of expanding insulating foam through the cut next to the boxes and foamed them in place, making sure that the foam stayed on the outside of the device box. They stayed put.

—*Bob Orlando and Paul Hoffman, Lafayette, Colo.*

CLAMP-NAILING

Pull nail into
stud when space is
too tight to swing a hammer.

Sometimes an electric outlet or switch box has to go between a
couple of studs that are so close together, there's no room for a
hammer or a drill bit. In this case, I reach for a C-clamp. As shown
here, a clamp can be used to squeeze a nail into the stud. For good
bearing, I use roofing nails during this operation. Occasionally, I
find it's necessary to drill holes in the side of the box for the nails.

—Dave Kohler, Clarks Summit, Pa.

FISHTAPE TARGET

In order to save ourselves some expensive plaster repairs, my
partner and I decided to route our new electrical wires directly
from the basement to the attic. We bored holes into the same
stud cavity from above and below, but we had a terrible time
trying to fish the new wire past the joist and the wall plates over
the basement.

Then we hit upon the idea of dropping a plastic bread bag tied to
a string from the attic into the stud bay. Once the bag was at the
bottom of the cavity, it made a big, easy-to-snag target that was
quickly caught with our fishtape and pulled into the basement.

—David Gelderloos, Boulder, Colo.

FISH-SNAKE HELPER

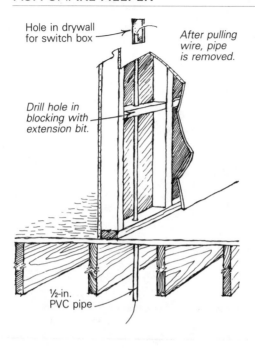

Hole in drywall
for switch box

After pulling
wire, pipe
is removed.

Drill hole in
blocking with
extension bit.

½-in.
PVC pipe

A fish snake makes it a lot easier to thread electric wire through walls, but sometimes the snake needs a little help. In my experience, that help is a length of plastic pipe. For example, let's say a piece of blocking lies between the basement ceiling and a new switch box, as shown in the drawing above. After drilling a 1-in. dia. hole in the blocking with an extension bit, I can probe around with the pipe until it passes through the hole. The pipe then becomes a temporary conduit through which the wire can pass.

I found another application for this temporary conduit approach when I had to fish a wire through a dropped ceiling in a bathroom. The wire had to go around a recessed light, past a number of other wires, and over some plumbing lines. The snake wire kept getting deflected by these obstacles. For help I turned to a piece of flexible ¾-in. polybutylene tubing. By rotating the tubing and wiggling it around, I was able to push it past all the obstacles. I could easily tell when it hit the far wall. As I pushed it against the distant drywall, my helper located its exact position by the tapping sound it made. Then we cut a small hole, ran our wires and removed the tubing.

—*Larry Wilson, Uniondale, Pa.*

NEW LIFE FOR OLD HOSE

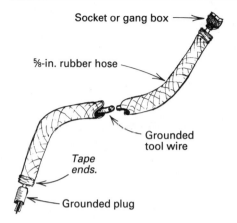

Socket or gang box →

⅝-in. rubber hose

Grounded
tool wire

Tape
ends.

Grounded plug

As a demolition contractor, I run through my fair share of equipment and tools—especially extension cords and rubber hose. Here's how I use one to help prolong the life of the other.

I make heavy-duty extension cords by pulling 14-ga. grounded tool wire through lengths of worn-out ⅝-in. rubber hose that would ordinarily be discarded. I pull the wire through using a fish tape and cut it to be about 3 in. longer than the hose to allow easy connection of the plug and the socket (drawing above). Once they're connected, I wrap the ends with waterproof tape.

Electrical accessories deserve special attention. Here are some safety precautions to follow in assembling this kind of extension cord. Make sure that hose sections are completely dry inside before you pull the wire through, and use only high-quality grounded wire of the correct gauge and equally high-quality plugs and sockets. Never submerge the cord and be sure to disassemble the sections periodically to check the condition of the wire for cuts or other signs of wear.

—*Scott N. Ayres, Bloomfield, N. J.*

HOT TIP

16-ga. copper
wire extension

Recently I had to repair the circuit board in my smoke alarm. For a job of this delicacy, I needed a soldering iron with a small-diameter tip, but all I had was a soldering gun with a fat, cumbersome electrode with a curved end. So I made a pointy-tip extension out of 16-ga. copper wire, as shown in the drawing above, and affixed it to the existing tip with twists of copper wire. The new tip allowed me to make precise patches to the tiny circuits.

—Joseph Kaye, Uniondale, N. Y.

THE COBRA LIGHT

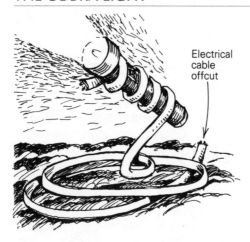

Electrical
cable
offcut

As an electrician, I spend time in crawl spaces running wires. For a quick trip, a flashlight is more convenient than a drop light— especially when space is tight. As shown above, I use a sheathed electrical cable offcut to wrap my flashlight into a standup fixture. This gives me another hand when I need it.

—Janet Scoll, Richmond, Calif.

A WORK LIGHT

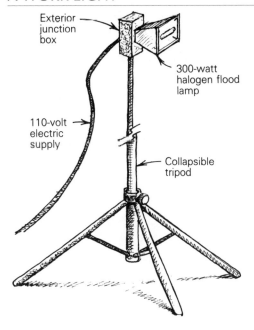

Exterior junction box

300-watt halogen flood lamp

110-volt electric supply

Collapsible tripod

I believe that you can t do good work if you can't see what you're doing, so I built the job light shown here to illuminate my work area. It's made up of a heavy-duty aluminum tripod (the kind used to support public address speakers) and an exterior 300-watt halogen flood lamp. I discovered that the knuckle at the back of the lamp fixture would fit into an exterior junction box, and that the junction box would slip over the top of the tripod shaft. The telescoping shaft extends to 72 in., and it rotates, allowing me to easily direct the light in any direction.

The tripod base is 3 ft. wide, and this generous span is the key to the success of this system. I've bumped into it many times, but not once have I knocked it over.

—Michael J. Geier, Hackensack, N. J.

ANOTHER WORK LIGHT

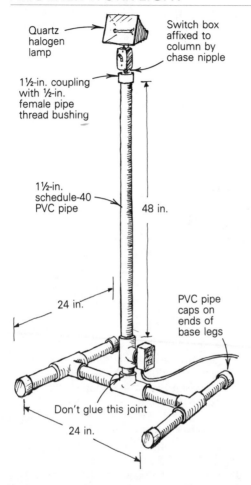

Quartz
halogen
lamp

Switch box
affixed to
column by
chase nipple

1½-in. coupling
with ½-in.
female pipe
thread bushing

1½-in.
schedule-40
PVC pipe

48 in.

24 in.

PVC pipe
caps on
ends of
base legs

Don't glue this joint

24 in.

For some time now I've been using light stands made of 1½-in.
Schedule 40 PVC pipe and fittings. A single 10-ft. section of pipe is
enough to make the stand shown above. To shed plenty of light on
my work, I use either a 300-watt or a 500-watt quartz halogen lamp
affixed to a metal outlet box with a switch in it. The box is attached
to the top of the stand by way of an electrical fitting known as a
chase nipple, and is screwed into a PVC bushing with ½-in. female
pipe threads. The bushing nests in a 1½-in. PVC coupling. The
bottom of the column has a PVC tee fitting with its own outlet and
cord. A 3-in. long piece of PVC pipe below the tee completes the
column portion of the stand. I glued all the joints in the column
and the base, except for the intersection between the two. This

makes it easier to break down the stand for transport. This light is equally useful for outdoor work, but if you plan to leave it outside you should use weatherproof boxes for the electrical fittings.

—Steve Harman, Okanogan, Wash.

BUZZLESS BULBS

A contractor friend of mine recently replaced some of my light switches with dimmers. I like the adjustable light levels, but I hated the buzzing noise made by the bulbs when they are dimmed. After some research, I discovered the solution—three-way bulbs. Their filament construction is such that they don't buzz when dimmed.

—Mary Jacek, Alameda, Calif.

10

QUESTIONS AND ANSWERS FROM BUILDERS

RECTIFYING A ROTTED GIRDER

The main support girder in a 100-year-old home I am renovating has some substantial rot. The girder is 6 in. by 8 in. by 14 ft. long; it's set in limestone foundation walls and has one support column in the middle. The support column was added later. The floor has sagged about 2 in., but I don't want to level it and introduce further complications. I have been given conflicting ideas and advice on replacing this girder, ranging from complete removal and replacement to bolting 6-in. C-channel steel on each side to strengthen it. Any ideas or advice?

—Greg Benson, Wayne, Ill.

George Nash, author of Renovating Old Houses *(The Taunton Press), replies:* Without knowing the framing layout or where the deterioration has occurred, it's difficult to offer a specific cure. But because masonry draws moisture, it's probable that the beam has rotted where it rests on its limestone supports. Also, noting that the mid-span support column is a later addition, I'd guess that the sag developed because the unsupported beam was too small for its load. So long as the present post isn't also rotting at its base, that problem, at least, has been cured. If jacking up the beam and shimming the post to remove the sag will crack upstairs wall plaster or require more than minor trimming of doors, you're probably wise to accept the dip in the floor as one of the charming idiosyncrasies of an old house. Otherwise, I'd recommend straightening out the sag as part of repairing the girder.

The need for replacement depends on how much and where the beam has rotted. Drive an ice pick, awl or screwdriver into the beam to find the depth of the rot. If at least two-thirds of the beam is still sound, it can be stiffened by sistering new wood to both sides of the beam with spikes of sufficient length to grab sound wood. The ends of the sistered timber must also rest on the limestone wall. If it proves too difficult to remove stones for this purpose, the new wood can be supported by a post under each end, tight to the wall. It's not elegant, but it works. This same method of posting will carry the weight of the beam if it has rotted at its wall support.

I'm assuming that the floor joists rest on top of the girder. Otherwise, it isn't possible to stiffen it by sistering. If this isn't the case, you'll have to add a new support timber underneath the existing girder and either remove or shorten the center post or add posts as needed. A new timber underneath will tend to straighten out the 2-in. sag, so shim or saw the timber to fit the existing contour of the girder.

If the beam has seriously deteriorated, complete replacement is probably the best strategy. Steel C-channel is expensive and isn't effective if the beam is rotted where it bears on the wall. Support the floor joists and/or any concentrated loads the girder carries with jacking timbers set under the joists on both sides of the girder. Raise these jacking timbers just enough to relieve the load on the girder. If joists or floorboards are nailed into the beam, cut the nails with a reciprocating saw before you try to remove the beam.

Before installing the new beam, soak the exposed faces of the old joists with a preservative, such as Cuprinol Green (Cuprinol Group, Div. of Sherwin Williams, 101 Prospect Ave., 15 Mid., Cleveland, Ohio 44115; 216-566-2000) to arrest further rot. Build up the new beam in place by nailing together four pressure-treated 2x8s. Jack the center of each plank up slightly to tension it before face-nailing it to the others. This prevents the new beam from sagging under load.

If floor joists are mortised into the girder, they should be sawn through and reattached with steel joist hangers. It's unlikely that the old joists will fit conventional hangers designed for modern dimensional lumber, so the joist hangers will have to be custom-made at a welding shop.

Adequate cross ventilation, such as that provided by operable basement windows, is necessary to retard the growth of rot-producing fungi throughout the cellar. High moisture levels or seasonal water infiltration also foster structural rot and must be prevented by good foundation drainage. Even nominally dry earth floors cause excessive moisture buildup; they should be covered with a plastic vapor barrier, an inch of protective sand and several inches of crushed pea gravel if a concrete floor is otherwise impractical.

REPLACING A SILL ON GRADE

I recently purchased a 130-year-old farmhouse in southern Ontario. Upon inspecting the foundation under an attached summer kitchen, I found that the 12x12 sill beam had been resting on the ground for all those years and is rotting. I suspect the joist ends that are mortised into the beam have also rotted.

I was intending to jack up the summer kitchen (a relatively light one-story frame structure), pour a concrete bond beam on grade and replace the sill beam and the joist ends. But how do I install jacks when there's no room under the joists? Also, do I need piers below the frost line under the bond beam? And if so, how do I dig out the space directly below the beam?

—Andrew Golbourne, Ont.

George Nash, author of Renovating Old Houses *(The Taunton Press) replies:* You don't mention what the soil is like under the farmhouse in question. Unless the house sits on well-drained sand or gravel, you'll need footings dug below the frost line. When founded on soil that is subject to frost heave, an entire building can move up and down a few inches each winter. But because your sill beam is continuous, this effect probably isn't very noticeable; however, freezing and thawing soil will take your bond beam along for the ride. I'm not sure from the drawing you supplied whether or not the opposite foundation wall extends below the frost line, either. It should.

Without inspecting the building myself, I can only offer a tentative solution. I'd excavate a trench along the length of the sill beam down to the frost line (4 ft. below grade is a safe bet). This should be 3 ft. or 4 ft. wide so that you can work comfortably in it. To hold the building, I'd spike a 2x10 ledger to the outside wall, directly into the wall studs. If possible, remove enough siding to accommodate the ledger. Set each jack on a wood pad in the bottom of the trench as close to the line of the sill as you can get it. Insert 4x4 posts between the jacks and the ledger (drawing, facing page). Try to keep these lifting posts at a 60° angle or steeper. The idea is to hold the wall, so apply only enough pressure to lift the sill barely off the stones it rests on. Do this at each end and every 8 ft. or so along the wall. I'm assuming that because this is a light stud-

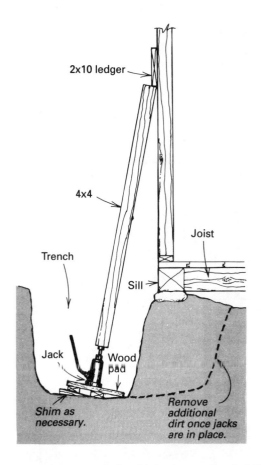

2x10 ledger

4x4

Joist

Trench

Sill

Jack

Wood pad

Shim as necessary.

Remove additional dirt once jacks are in place.

framed structure, it shouldn't take much to hold it in place. If it's a timber frame, lift on each corner and on any intermediate posts. Then come back with the backhoe and dig out under the sill and as far back under the joists as you can reach. This should give you enough room to slide a timber under the joists to support them temporarily while you remove the rotted sill.

Next, form and pour a concrete footing. Because you'll probably have a hard time finding a 12x12 timber for the new sill, I'd double up two 6x12s (flat) or some other combination to give the necessary support above. Jack this sill up from the footing tight to the flooring (drawing, p. 164). If you use steel cellar posts (you might have to cut them down a little) that have built-in jack screws, you can leave them in place and pour the new wall around them. I'd sister new joists alongside the old ones (this is why you dig under the sill with the backhoe) and secure them to the back of the new sill with metal joist hangers.

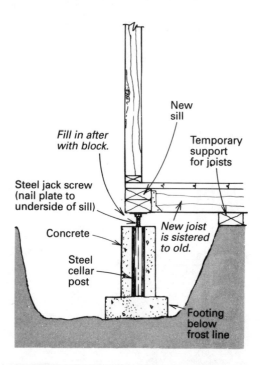

Fill in after with block.

New sill

Temporary support for joists

Steel jack screw (nail plate to underside of sill)

Concrete

New joist is sistered to old.

Steel cellar post

Footing below frost line

I'd pour about 32 in. of concrete foundation wall (borrow or rent some form panels or build them from plywood) and then cap it off with a course or two of concrete block. Unless the soil is expansive clay, you can build the entire wall from block. Fill in any gap between block and sill with pressure-treated wood and a layer of insulating compressible foam. You may have to raise the sill slightly higher to do this.

An alternative is to construct the foundation wall before you replace the sill. The sistered joists extend to the new wall and onto a conventional mudsill and rim joist built up with treated dimension lumber. In either case, don't forget to provide some sort of access opening (if there isn't one already) to get under the crawl space. Provide proper drainage when you backfill the trench and be sure that the finish grade is at least 8 in. below the bottom of the sill.

BATTLING A BULGE IN A FOUNDATION WALL

About three years ago I built a three-story, 3,000-sq. ft. home. I sublet the foundation to a local contractor who constructed 8x10x16 concrete-block walls 12 courses high on 12-in. by 24-in. footings. After installing a 4-in. French drain covered by 4 ft. of gravel, the backfill was placed (following rough framing) to within 1 ft. of the top of the block. Two exterior concrete-block pilasters were built along the 36-ft. long walls to bolster the structure.

Recently I have noticed a horizontal crack, at midheight, running along this foundation wall. The center of the wall, at midspan, bulges out about ¼ in. Drainage from the foundation is good with little or no moisture inside the basement.

What can I do to stop the bulging of the wall and its ultimate failure?

—Paul Kane, Pittsburgh, Pa.

Dick Kreh, master mason and author in Frederick, Md., replies: The crack you describe is the result of great pressure along the wall at midspan. First, a wall as high as you described and filled to within 1 ft. of the top should have been built of 8x12x16 block instead of 8x10x16 block. That would have helped withstand the pressure. Too late now. The two exterior pilasters should have been built on the inside, where they would have braced the wall more effectively. The pilasters are probably not restraining anything.

I suspect that the heavy equipment used during backfill caused a hairline crack that has now widened from moisture and earth settlement. I don't know what type of equipment the contractor used, but the pressure created from pushing earth against a wall with a 30-ton bulldozer is enormous. It would have been wise to brace the inside of the foundation wall with framing lumber until the earth had time to settle. From the drainage described, I don't think that water is causing the crack; however, you should make sure that the downspouts on your rain gutters are not emptying into the problem area.

If you can see a crack on the inside of the wall, the wall must be cracked on the exterior, too. To repair the wall, I'd excavate the earth on the exterior down to and a little below the crack. I will bet that this alone will let the wall come back to its original position. Using a joint chisel, cut out the cracked bed joint (the horizontal mortar) approximately ¾ in. to 1 in. deep. Wet down and brush the chiseled-out joint and repoint it with Type-S masonry cement mortar, available at any masonry supply. I would do my repointing with a flat metal tool called a slicker, which forces the mortar

Section through wall and pilaster

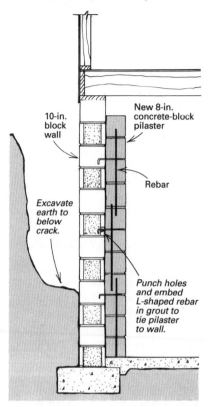

10-in. block wall

New 8-in. concrete-block pilaster

Rebar

Excavate earth to below crack.

Punch holes and embed L-shaped rebar in grout to tie pilaster to wall.

tightly into the joint. Let it set a day, then apply a coating of foundation waterproof tar or mastic as an insurance sealer.

Inside the basement, cut out the cracked bed joint and repoint it the same way. I would then build two pilasters of 8x8x16 block, 24 in. wide (one block and a half), laid out at one-third and two-thirds the length of the wall (drawing above). As you build the pilasters, cut holes in the foundation wall at three heights. When the pilaster meets a hole, fill the hole with mortar and insert an L-shaped piece of rebar. Lay the straight end of the rebar on the block in the mortar before laying the next course of block to tie the pilaster to the wall for extra protection. Inserting rebar vertically in the hollow cores of the pilaster blocks as they are laid up and filling the pilasters with mortar strengthens them considerably. Allow the pilasters to cure at least a week, then gently fill the earth back against the outside wall, preferably with a shovel, and tamp lightly. The entire operation is labor-intensive, but I think it's the only method that will work.

READERS REPLY

Dick Kreh's solution in "Battling a bulge in a foundation wall" (see p. 165) will certainly fix the problem, but it will be difficult and expensive. Excavating the soil down to the crack may or may not allow the wall to return to its original position. However, if you don't know where this crack is, you won't find it from the earth side. The crack described in this problem is most likely a flexural crack, not a settlement crack. A settlement crack shows on both sides of a wall, but with a flexural crack the earth side of the wall will be tight.

Unless a foundation drain is required, there's really no reason to dig. A simpler, less expensive solution involves using steel I-beams instead of masonry pilasters (drawing below). The steel I-beam can be fabricated in a shop, and the home owner can bolt it in place himself. It gets bolted to the floor, to the block wall and to a floor joist above with steel bolt plates. Eight plates are welded to the

Support for cracked block wall

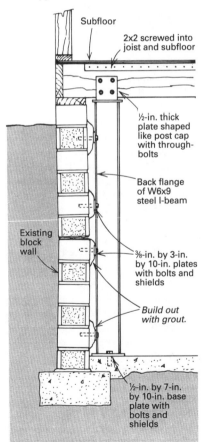

Subfloor

2x2 screwed into joist and subfloor

½-in. thick plate shaped like post cap with through-bolts

Back flange of W6x9 steel I-beam

Existing block wall

⅜-in. by 3-in. by 10-in. plates with bolts and shields

Build out with grout.

½-in. by 7-in. by 10-in. base plate with bolts and shields

I-beam: one on the bottom, four along the back flange and three on the top. These top plates should be fabricated like a post cap to fit around the floor joist. The length of the I-beam equals the height from the slab floor to the floor joist minus about an inch so that there's some room to play with.

The connection between the floor joist and the subfloor will have to be beefed up, which will help transfer the load from the I-beam to the floor diaphragm. Screwing a 2x2 onto either side of the floor joist and into the subfloor every 3 in. o. c. will do the job.

—W. F. Palmer, Jr., P. E., Chamblee, Ga.

In response to Dick Kreh's answer, although the existing pilasters were probably undersized and/or spread too far apart to prevent a flexural crack, the mason was correct to have built exterior block pilasters. Unreinforced masonry pilasters work well outside a basement wall, but reinforced masonry pilasters are best located inside the basement. The reasons why this is true boil down to the relative virtues of masonry and steel. Masonry is fairly strong in compression (meaning it's being squeezed); it can support a great deal of weight before it gets crushed. On the other hand, masonry is lousy in tension (meaning it's being stretched). The bond between mortar and masonry can handle very little stretching before it cracks. Once it cracks, the tensile strength of the system is nil.

By comparison, steel is terrific in tension. For example, the stress that a standard steel reinforcing bar can withstand before failure is about 125 times that which typical masonry mortar can resist.

Ignoring for the moment the vertical load basement walls usually support, a basement wall acts like a wide beam—spanning from the basement slab to the subfloor above—to resist the lateral pressure of earth and water pushing in from the outside. The inside face of the basement wall is in tension. The outside face is in compression. Somewhere in between, a vertical plane exists that's neither in compression nor in tension: It's not being stressed at all.

The location of this plane is called the neutral axis, and it coincides with the center of gravity (or mass) in any wall. In a block wall without pilasters, the neutral axis is right down the middle. The stress at any point in the wall is directly proportional to its distance from the neutral axis. Add pilasters, and the wall's neutral axis moves closer to the pilaster side. Even though the neutral axis moves away from the tension side when pilasters are added on the outside, the increased stiffness provided by the pilasters more than offsets this increased distance from tension face to neutral axis. The net result is a decrease in the tension stress, which masonry resists so poorly, along with an increase in compressive stress, which the masonry can handle easily. Reinforced pilasters, on the other hand, should go on the inside of the wall, as Mr. Kreh suggests.

—Christopher F. DeBlois, P. E., Chamblee, Ga.

BUILDING A WALL ON AN EXISTING SLAB

I have a concrete patio in the corner of my L-shaped house. I plan to convert the patio to an enclosed sunroom. The patio slab extends 4 in. beyond the foundation, and I want to keep the new walls in the same plane as the existing walls. But I have two small children, and I'm afraid that angling flashing over the 4-in. lip would be dangerous, not to mention unsightly. How can I construct the sill to eliminate water problems?

—John Bennett, Grover, Mo.

Fine Homebuilding's contributing editor, Scott McBride, replies:
You've got several options. If you don't mind leaving the slab as an exposed ledge, you can simply bolt a curb onto the slab that will reach up above the lower edge of the new siding. The curb can be made from masonry or from wood.

If you choose a masonry curb (drawing, p. 170), check with a mason's supply yard for the CMUs (concrete masonry units) that suit your purpose. A yard near me stocks a 3½-in. by 9½-in. cast lintel that might serve nicely. The 3½-in. width will work with a 2x4 wall on top. These lintels are available in lengths up to 10 ft.

Your local building code may call for the wall to be anchored to the slab with threaded rods. If this is the case, drill through the curb and into the slab with a roto hammer every few feet. Remove the curb and clean the slab thoroughly with a wire brush. Then fill the holes in the slab with hydraulic cement.

Hydraulic cement is extremely tenacious and waterproof. It's also expensive, difficult to work with and sets up fast like plaster of Paris, so have everything ready before you mix. Set the threaded rods into the cement-filled holes; make sure the rods are long enough to pass through a wood sill that will sit on the curb. Then place a bed of rich mortar (one part masonry cement to two parts sand) and slip the curb over the rods into the mortar bed. Rake out the bed joint (where the curb meets the slab) ¾ in. deep on the exterior side. When the mortar has cured, parge the new curb with a thin (⅛-in. thick) layer of hydraulic cement. For positive drainage, build up a sloping surface on the ledge. While the parging cures, cover it with a tarp for a few days; exposure to direct sunlight weakens cement by causing it to cure too quickly.

The easiest route of all would be to bolt a pressure-treated 4x4 to the slab to act as a curb (either with threaded rods or lag screws and lead shields) as shown in the drawing on p. 171. Drying the 4x4s for a few months before you install them will reduce the shrinkage factor. To seal between the wood and the slab, you could use a thick bed of premium caulk, but a better alternative

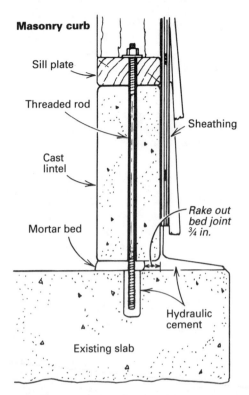

Masonry curb

Sill plate

Threaded rod

Sheathing

Cast
lintel

*Rake out
bed joint
¾ in.*

Mortar bed

Hydraulic
cement

Existing slab

would be an extruded EPDM sill gasket. Resource Conservation
Technology (2633 North Calvert St., Baltimore, Md. 21218; 410-366-
1146) makes them in three widths for 2-in., 4-in. and 6-in. frame
walls. An EPDM sill gasket will expand somewhat to fill gaps as the
4x4 shrinks. You can parge the exposed face of the 4x4 to match
the foundation by covering it first with galvanized wire lath.

If you don't want a 4-in. ledge, you can cut the concrete slab
back to align it with the rest of the foundation. You can do most of
the cutting with a diamond blade in a cutoff saw. (For more on this
tool, see *FHB* #62, 80-84.) Snap lines and make repeated passes
with the saw until you reach the full depth of the blade. For each
pass, lower the blade so that only its diamond-coated section does
the cutting. A deeper cut causes the blade to overheat. Use a
straightedge to guide the saw. After reaching the maximum depth—
about 5 in. with a 14-in. blade—bust off the waste with a sledge. If
you can't cut all the way to the end of the slab because the saw
runs into the house, use a roto hammer to finish the cut. Cutoff
saws and roto hammers are standard fare at rental yards.

The diamond-cut edge will be nice and smooth, but you should
patch the rough ends of the cut with a concrete topping mix. If the

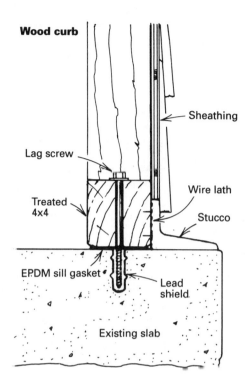

Wood curb

Sheathing

Lag screw

Treated 4x4

Wire lath

Stucco

E.P.D.M sill gasket

Lead shield

Existing slab

slab is high enough from the ground, you won't need a curb; you can simply begin framing. If not, you can build your new wall using one of the methods I just described.

STABILIZING A CRACKED SLAB

My one-year-old house was built on a slab, a common practice in northwest Florida. During construction, the slab developed small stress cracks. These minor cracks have only caused problems in one bathroom, where a dozen floor tiles have cracked. We intend to replace the cracked tiles, but are concerned that the replacements could crack, too.

How can I tell when the cracking will cease? How, without doing major foundation work, can I strengthen the slab against further cracking? In case the slab is still unstable, is there a process for removing and replacing the broken tiles that will restrict further cracking to the grout lines?

—*James L. Harvey, Niceville, Fla.*

Construction consultant Alvin Sacks replies: The only sure way to determine whether a slab is through cracking is to wait it out. How long to wait is up to you, but cracks can keep occurring for years. The problem could be caused by a poorly placed slab, or by underlying soil that wasn't properly compacted. In either case, the most thorough fix is to remove that portion of the slab that has major cracks.

The next step is to place a new 4-in. thick slab. Add reinforcing bars for tensile strength, and wire mesh for crack resistance. Where the old slab meets the new, they should be doweled together with reinforcing bars. An excellent source of information on concrete reinforcement is the 71-page *Residential Concrete,* prepared by the National Assocation of Home Builders. It's available for $12 (plus $3 shipping) from the NAHB bookstore (15th & M Streets N. W., Washington, D. C. 20005; 800-223-2665).

Before re-laying the tiles, cover the slab with a membrane (either a polyethylene or bituminous sheet). The membrane will serve as a water retarder as well as a slipsheet to permit movement between the slab and the tiles. Lay the tiles in a 2-in. thick mud base. To minimize future cracking, use a grout-matching silicone or polyurethane sealant as a grout substitute in strategic areas. Such areas include the intersections of tile with walls and fixtures, as well as every 4 ft. in both directions throughout the field. The membrane and the soft joints will protect the tile by absorbing a certain amount of slab movement.

If you want to take a chance with the existing slab and just replace the broken tiles, I know of only one way to restrict future cracking. First, fill the cracks with a liquid water-retardant membrane (your tile supplier can recommend a specific product). This will protect the tile from water seepage from below. Then lay the tile directly over the concrete. Alternate the field joints between grout and a matching sealant. Use matching sealant around the edges of the patch. You may not get an exact match between the new grout and the old, but chances are the match will improve with age.

FLASHING A WALL

I plan to enclose a covered porch (drawing, facing page), and I am concerned about maintaining a waterproof seal between the bottom plate of the new exterior wall and the existing concrete pad. Also, there are two valleys that channel water to the existing pad, and during heavy Texas downpours, a tremendous amount of water hits this pad and will splash on my new wall. I want to protect the

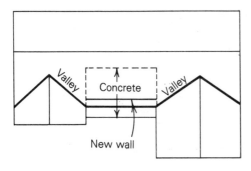

insulation and hardboard siding from water damage. I
would like your recommendations for caulking and
flashing this wall.
—Dan McCubbin, Bedford, Tex.

*Stephen D. Dahlin, a consulting engineer specializing in building
technology, replies:* The first thing to do is divert those Texas
downpours into a gutter system, with downspouts letting out a
good distance away from the slab. To size the gutters and
downspouts, you can use the recipe in *Copper and Common Sense,*
a book published by Revere Copper Products, Inc. (P. O. Box 300,
Rome, N. Y. 13440; $15).

The second task, waterproofing the base of the wall, is not as
straightforward. First, consider cutting off and removing the
portion of the slab that would extend beyond the new wall. If that
proves too expensive, is it possible to position your wall at the edge
of the slab? If you can do either of these, you'll have an easier time
waterproofing the wall.

If you have to locate the wall away from the edge of the slab, I
recommend the construction shown in the drawing at the top of
p. 174. The bottom plate of the wall sits on a strip of ½-in. pressure-
treated plywood and is waterproofed with metal flashing, which is
doubly sealed to the concrete slab. The metal starts several inches
up the wall, extends down and returns under the plate. If your wall
will have a door in it, it will need a curb under the threshold, with
the flashing unbroken, covering the sill and jambs (drawing,
bottom, p. 174). This means using a metal that can be soldered—
copper or stainless steel. If there's no door, soldering will not be
necessary and aluminum would be another option. Copper should
be 16 oz. minimum; aluminum should be 20 ga.

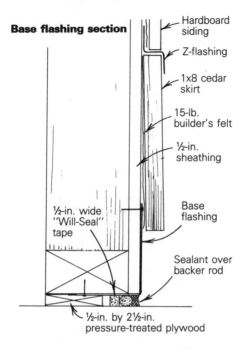

Base flashing section

Hardboard siding

Z-flashing

1x8 cedar skirt

15-lb. builder's felt

½-in. sheathing

½-in. wide "Will-Seal" tape

Base flashing

Sealant over backer rod

½-in. by 2½-in. pressure-treated plywood

Your flashing should have an expansion joint near the center of the wall (but not in a doorway) that allows for ³⁄₈-in. of movement (drawing, facing page). Fill the overlaps in the joint with sealant. Use nails that are compatible with the metal: copper nails for copper metal, stainless steel in kind, and stainless or hot-dipped

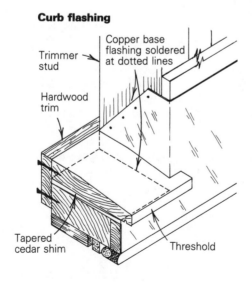

Curb flashing

Copper base flashing soldered at dotted lines

Trimmer stud

Hardwood trim

Tapered cedar shim

Threshold

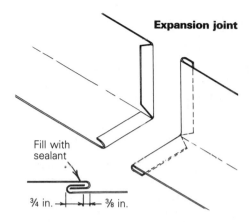

Expansion joint

Fill with sealant

¾ in. — ⅜ in.

galvanized steel for aluminum. Maze Nails (W. H. Maze Co., 100 Church St., Peru, Il. 61354) is one good source of such nails.

The primary seal is a bead of sealant against a backer rod. A high-grade sealant is in order; I'd use a polyurethane caulk such as Sikaflex 1A (Sika Corporation, 92 River St. Montpelier, Vt. 05602) or Dymonic (Tremco, Inc., 3735 Green Rd., Cleveland, Ohio 44122). Be sure to follow all the manufacturer's recommendations for cleaning and priming the surfaces and for applying and tooling the sealant.

The secondary seal is an expanding bitumen-laden foam tape, such as Will-Seal (Illbruck Inc., 3800 Washington Ave., North, Minneapolis, Minn. 55412). This tape should provide a long-term seal while the caulking, which will protect the tape from weathering, may need replacing in as little as seven years.

Prepare the surfaces and install the tape before installing the bottom plate and the metal flashing. Insert a ¾-in. dia. closed-cell polyethylene backer rod into the space between the plate and the concrete pad. Push it far enough so that the sealant will be ¼ in. to ⅜ in. thick at its thinnest point. Apply the sealant by pushing the sealant ahead of the nozzle to fill the joint solidly. Tool the bead to a slightly concave shape, using a dry, spoon-shaped trowel or a small spoon, frequently cleaning the tool with the recommended solvent.

At the bottom of the wall, I have shown a 1x8 cedar skirtboard whose bottom edge overlaps the metal flashing and aligns with the siding on adjacent walls. Although it introduces an odd design feature, its weather-resisting function is important. Alternatively, a course or two of beveled wood siding might better please the eye. Either way, your hardboard siding will be farther away from splashing water. Use an aluminum Z-flashing between the hardboard and the skirt. Use 15-lb. builder's felt over the sheathing.

FROST WALL QUESTIONED

Building in our area requires a 4-ft. deep frost wall to avoid frost heaving. My problem involves the construction of a playroom and porch in what was an unheated, attached garage. Where the old garage doors had been, porch railing and balusters will be installed. Four feet behind the railing, a new wall will be constructed to close off the playroom. This new exterior wall will rest on the old garage floor (drawing below) and not over a frost wall. Should the concrete floor be broken up and a new frost wall poured under the location of the new wall? The garage floor has never moved in the ten years that it has been there, and the garage was never heated.

—Jack Howes, Peterborough, N. H.

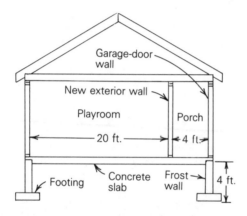

Architect Thomas Reardon of Northboro, Mass., replies: The existing frost wall under your garage-door wall is structurally adequate to continue carrying your garage roof. If the garage-door openings are to be increased, however, replace the headers with larger ones to carry the roof loads. If you remove the existing garage-door wall completely, the new recessed exterior wall will become a bearing wall and will require a 4-ft. frost wall below.

Another non-structural but problematic condition created by your conversion is an unheated and heated space using the same concrete slab as a common floor. Without some kind of thermal break at the new exterior wall, the existing floor slab will act like a thermal wick during winter, conducting heat out of your conditioned space. You might want to consider insulating the floor in the playroom or else adding an expansion joint to the slab as a thermal break.

WINTER FOUNDATION PROTECTION

I am building two houses, and for reasons of time and finances their basements must be completed during the summer with an intervening winter before they are framed and finished. The houses are located in Schenectady, N. Y., and Mt. Vernon, Ohio. If the basement walls are capped with the first-floor deck, what other measures should be taken, such as sealing and heating, in order to protect the new foundations from winter damage?

—Walter Kaufmann, Mount Vernon, Ohio

Consulting editor Tom Law replies: The chief enemy is frost heaving. Treat an unattended basement like any other house to protect it from freezing below the footings. The upper floor should be on and water shed with plastic or builder's felt. Drain tile should be in place and walls backfilled, with finish grade sloping away from the wall in every direction. Close off any openings and monitor the temperature. In extreme cold, introduce heat if required. Even though the earth mass outside is sufficient to prevent freezing at footing level, the floor and walls can get cold enough to cause the earth below the footings to freeze.

SUPPORTING EXTERIOR STAIRS

When building exterior stairs up to a deck, we typically use open stringers with a pair of 2x6 treads and a rise of about 8 in. With this constructions, the top of the stringer barely meets a 2x10 skirtboard, (drawing below), which we use to cover the rim joist. I've tries different methods of attaching stringer without ever being satisfied. The weight of the stairs is exerted against the face of the skirt, and the typical 1½-in. overlap is probably adequate structurally. But the

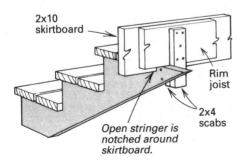

2x10 skirtboard

Rim joist

2x4 scabs

Open stringer is notched around skirtboard.

connection must be strong enough to keep the stringer from shifting or warping.

What I'm doing now is notching the stringer and extending it behind the skirtboard, then nailing through the side of the stringer into a pair of 2x4 scraps scabbed vertically to the back of the rim joist. This works, but is awkward and looks cobbled. Can you suggest a more secure way of attaching the stringer?

—Brian Zaintz, Ont.

Scott McBride, contributing editor of Fine Homebuilding, *replies:* I too have hung porch stair stringers on 2x4 scab blocks, and I agree that it seem a rather clunky solution. There are a couple of alternatives.

My first preference would be using a closed stringer (top drawing below). The treads can be supported either on cleats or in routed housings. The closed stringer will bear fully on the rim joist, with plenty of area for toenailing. The closed stringer affords purchase for newels and balusters, as well. Above all, it's much

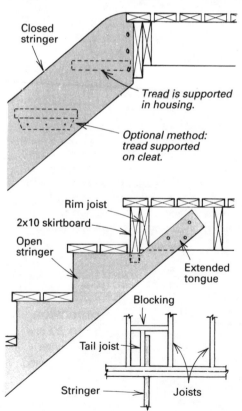

Closed stringer

Tread is supported in housing.

Optional method: tread supported on cleat.

Rim joist

2x10 skirtboard

Open stringer

Extended tongue

Blocking

Tail joist

Stringer

Joists

stronger than an open stringer of the same dimension—an important consideration for stairs with considerable span. To lighten the appearance of a closed-stringer stair, I use a jigsaw to radius the top corner where it levels off to meet the deck.

If you prefer an open stringer, support it by extending it up into the deck framing (bottom drawing, facing page). This will require longer stringer stock, but will avoid that cobbled look. Extend only the lower third of the stringer's width, in a 4-in. wide tongue. The strength is in the unbroken lower third of the stringer's width anyway. You may need to notch the skirt a little to avoid paring down the tongue of the stringer too much.

If the deck joists run parallel to the stringers, you might be able to coordinate joist layout with stringer location. If a stringer falls in the middle of a bay, head off the bay with blocking and frame a short tail joist back from the blocking to the rim joist to pick up the stringer. Nailing in such narrow bays is a pain, so install the stringers from above before you lay the decking.

If the joists run perpendicular to the stringers, install backing across the first bay to pick up the extended tongue of the stringers.

LATERAL SUPPORT FOR WOOD BEAMS

At what point must a wood beam be braced against lateral forces? Does it depend on the load or the beam's cross section? I'm also curious about whether the rules for enclosed headers differ from those for exposed beams.

—Jim Warner, Fairfield, Ohio

Christopher DeBlois, a structural engineer from Atlanta, Ga., replies: How much lateral support a wood beam requires will depend on its depth-to-width ratio. A beam that's less than twice as deep as it is wide requires no special restraint. But one that's more than six times as deep as it is wide must have its top and bottom edges held in line along its full length—otherwise it will be unstable. This principle is easily demonstrated with a yardstick. If you lay the yardstick flat on two end supports and press down on the middle of the span, all it does is sag. But holding it on edge and pushing down on the middle will make the yardstick sag and twist.

You obviously don't want a structural member twisting in an unstable manner under a real load. Fortunately, this usually isn't a problem for typical in-wall headers. The top edge of the header is usually fastened to a top plate or to floor joists, and the ends are nailed or bolted to vertical framing members. This provides lateral support for the header along its entire length. For an exposed girder carrying only one or two beams, however, the threat of

Lateral Support Rules for Wood Beams	
Depth to width ratio	Rule
2:1 or less	No lateral support required.
3:1 to 4:1	Support the ends with bridging or joist hangers, or nail or bolt the ends to other framing members.
5:1	Hold one edge in line for its entire length.
6:1	Three alternatives: 1) support every 8 ft. with solid blocking or cross bracing, 2) hold both edges in line along their entire length, or 3) hold the compression edge in line (with subflooring or sheathing, for example) and bolt or nail the ends.
7:1	Hold both edges in line for their entire length.

twisting can be very real. The chart above gives general rules for lateral support. If you're at all concerned about the stability of a beam or header, have it checked by an engineer.

ON BEEFING UP JOISTS

Does a 2x6, 2x8 or 2x10 joist sandwiched between two lengths of ripped ¾-in. plywood have the same load capacity as a joist scabbed to a new member of the same size?

—Jack Kelbish, Virginia Beach, Va.

Daniel Brown, a structural engineer in Tacoma, Wash., replies:
There are two drawbacks to the use of plywood. Assuming that your joists are over 8 ft. long, the 8-ft. lengths of the plywood strips mean that their stiffening effect is less than if the plywood ran the full length of the joist. Second, because the plywood veneers are cross-laminated, only about half of the wood grain is in the direction of stress (in contrast, the veneers in the flanges of a wood I-beam all run in the same direction). This makes a ¾-in. thickness of plywood equal in strength to only ⅜ in. of solid wood in this

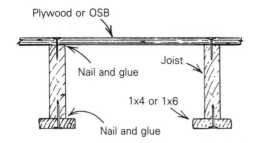

Plywood or OSB

Nail and glue

Joist

1x4 or 1x6

Nail and glue

application. So 1x8s would be much more effective than plywood strips; all of their grain is in the "right" direction and you can probably find boards that are long enough to be used in one piece.

There is a third option for stiffening springy joists (drawing above). Boards, 1x4 or larger, can be attached to the bottom edge of the joist with nails and construction adhesive. This creates a flange like that on an I-beam, adding considerable stiffness. To be most effective, the boards should extend from support to support, but it's acceptable for them to be short by 3 in. or 4 in. at each end. Gluing the subfloor to the top of the joists also improves stiffness.

REMEDY FOR A RACKED GARAGE

I own a garage that is now 18 ft. wide and 65 ft. long (yes, 65 ft. long). Originally only 22 ft. long, an addition was put on by the previous owners. The walls and roof trusses are framed with 2x4s. The 16-ft. garage-door opening is spanned by a doubled 2x10 header and flanked by 4x4s.

Several months ago I noticed the garage-door opening was out of square by about 3½ in. when I measured the diagonals. I have since built a wall dividing the structure about 22 ft. from the front, where the original back wall would have been. I did so to get more effectiveness from my woodstove, and I hoped to stop the building from racking.

On rechecking the door opening, however, I found it to be 5½ in. out of square. How can I get the building to stand straight, and how can I keep it there without adding a slew of posts that would greatly cramp my shop space?

—Frederick Mayo, Somerset, Mass.

William E. Burdick, a retired officer of the Navy Civil Engineer Corps, replies: Light-frame garages typically do not have the stability created by the continuously sheathed diaphragms of floors, walls and roofs, and are thus vulnerable to distortion, especially racking.

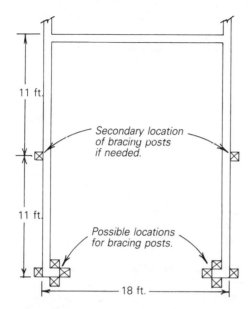

11 ft.

Secondary location
of bracing posts
if needed.

11 ft.

Possible locations
for bracing posts.

18 ft.

There are many ways to add bracing. As you imply, the trick is to select the easiest and cheapest effective method.

The strength of the existing structure and its connections is probably limited, so I won't recommend methods that depend on existing pieces of the garage. Such methods might include steel rod X-bracing seen in pre-engineered steel buildings, or wooden knee braces. These techniques, by tensioning and leverage, could cause stresses that overpower the structure they are intended to support, breaking connections and buckling posts and plates.

I suggest using the technique found in pole-barn construction. At two, or possibly four points, I would firmly plant timber posts in concrete and bolt them to the structure. The strength of this bracing is independent of the building.

Regardless of the solution used, the first step will be to pull things back to plumb. The only practical method I know is to use a ratchet come-along, connected on the diagonal between the top of one wall and the bottom of the opposite wall. When the garage is plumb, brace the door opening with temporary diagonals, say a 12-ft. 2x6 at each end, from the bottom of the 4x4 posts to the 2x10 header.

I would use, at the minimum, 4x6 pressure-treated posts—6x6s if the price differential is insignificant. Sometimes cutoffs from old and new power-line poles are available for the taking at the power-company yard. These would be suitable.

For major structures, sophisticated calculations are needed to determine the size posts and the depth of their embedment. Your requirement is at the minimal scale of load, so we'll keep it simple.

I suggest posts 12 ft. long, buried 4 ft. deep if possible, but at least 3 ft. deep. Try to keep the diameter of the post hole as small as possible. Brace the post plumb in the hole and backfill with concrete. I've had success using Sakrete or a dry site mix, well rammed into the hole, without adding water. The mixture is then hydrated by absorbing ground moisture.

The posts can be inside the garage if there is no floor slab, outside if there is. The drawing on the facing page shows some possible locations. After waiting four to eight days for the concrete to develop strength, bolt each post at top and near ground level to the 4x4s flanking your door opening, and remove the temporary bracing. Bolt them directly to the walls if footings permit; use spacer blocks if not. If you can't locate bolts that are long enough, a machine shop can readily thread some ⅝-in. dia. hot-rolled stock for the few bolts needed.

In addition to a post at each side of the door, you may wish to add additional pairs of posts if distortion is evident. Spacing at not less than 11 ft. (the halfway point of the garage) or at splices of the upper plate should be sufficient.

READERS REPLY
Editor's note: Below and on p. 184 are three alternative solutions to Frederick Mayo's racked garage.

My approach would be to install bracing diagonally across the ceiling (corner to corner). Two by sixes should be adequate for the braces. Securely tie the braces to the wall frames near the corners, and use a lap joint where they cross each other. Because each brace will need to be made from two boards to get the length you'll need, join them with a lap joint also. Bolt and nail all lap joints for strength. Nail the braces to the bottom chords of the trusses (which are acting as ceiling joists).

With the braces in place, any tendency to rack at the door opening will produce a racking force in the side walls. Since the side walls are braced against racking, the resulting force will attempt to lift a corner of the structure. Thus the building should be securely anchored to the foundation. Although a lifting force could separate the structure at a corner, if the framing and sheathing were sound it would require a force much larger than is likely ever to occur.

—*Bob Jenkinson, Ignacio, Calif.*

Use a ratchet-type come-along to return the door opening 2 in. past square as measured across the diagonals. Remove the siding on the garage-door wall. Glue and nail ½-in. plywood vertically on both the inside and outside faces of the wall; use 10d nails 12 in. o. c. When you cut out the excess plywood back to the door opening, you will have an L-shaped sandwich on each side of the door. When the come-along is released the door opening should spring back to square.

—Mark Gentry, Auburn, Calif.

After returning the structure to a plumb condition, stiffen the three main walls by transforming them into shear walls. This can be done by nailing ⅜-in. plywood, with fully blocked edges, to the inside surface of the walls. Minimum nailing should be 8d common nails spaced 6 in. o. c. along plywood edges, and 12 in. o. c. along intermediate studs. Additionally, the ceiling should be transformed into a structural diaphragm. This can be done by nailing ⅜-in. plywood, with fully blocked edges, to the ceiling. Minimum nailing should be as above.

In order for this method to work properly, it is imperative that the shear walls and ceiling diaphragm be properly tied together along their adjacent boundaries. This can be done by providing metal connectors (e. g., Simpson Strong Tie A34 connectors spaced at 18 in. o. c.) in the corners between the ceiling and walls. It is important to note that the nails used to secure the connectors must pass into solid framing behind the plywood sheathing.

—Paul Shanta, Shanta Associates,
Consulting Engineers, Hayward, Calif.

SPANNING 19 FT.

I am planning a second-story addition over my woodworking shop, and I don't want any support post in my way on the first floor. I want to make box-beam floor joists. What size should I make the box beams to span the 19-ft. width of the addition?

—Gary McClain, Darden, Tenn.

Christopher F. DeBlois, a structural engineer in Atlanta, Ga., replies:
A 19-ft. span carrying residential loads (10 lb. per sq. ft. dead load plus 40 lb. per sq. ft. live load) does seem to call for unusual framing. My first thoughts ran to wood I-beams, laminated veneer lumber or parallel-chord wood floor trusses. But it turns out that there is no need for your box beams or for any of these approaches. In fact, ordinary #2 Southern yellow pine 2x12s, with a

maximum moisture content of 19% and spaced 16 in. o. c., have plenty of strength to span 19 ft. (You should check this, of course, against your local codes.)

My primary concern with 2x12s at 16 in. o. c. is not their load capacity but their deflection and vibration behavior. Strength considerations limit this system to a 21-ft. span, but for a maximum deflection under live loads of $\frac{1}{360}$ of the span length (about $\frac{5}{8}$ in.), these joists can span only 19 ft. 11 in. To increase the stiffness of the system and thereby reduce both deflection and vibrations, follow the guidelines of the American Plywood Association (APA, P. O. Box 11700, Tacoma, Wash, 98411-0700; 206-565-6600) for its "Glued-Floor System." The system is described in the APA's publication *Design/Construction Guide: Residential & Commercial* ($2).

The glued-floor system doesn't increase the strength of your joists; the 2x12s still do all of the work. But it will increase the stiffness. Using subfloor panels rated 16 in. o. c. or 20 in. o. c., the allowable span jumps to 20 ft. 11 in. That extra distance may not sound like much, but it represents a 25% bonus in the floor stiffness.

The additional work involved to comply with the APA's recommendations is not all that great and yields a large benefit. You already need the 2x12s to span the 19 ft., and you're going to need a subloor, so the only extra costs are for the correct glue and the labor involved in gluing the floor properly. The system requires all joints in the subfloor panels to be glued with a glue that conforms to APA Specification AFG-01. If the label on a container of glue states that it conforms to this specification, then it is acceptable for use with the glued-floor system. Joints that run perpendicular to the joists must be either glued tongue-and-groove joints or square-edge-panel joints glued to blocking below.

The APA guide also discusses proper nailing. If your floor must also act as a horizontal diaphragm to handle wind or earthquake loads, more nailing may be required. Check with a structural engineer if there is any question.

READERS REPLY
I had a problem similar to Gary McClain's (see facing page): I had to support an open patio deck over my 20-ft. by 21-ft. wood shop without interior columns. We spanned 21 ft. using steel joists (#925SJ16, which means 9¼ in. deep by 1¾ in. flange by 16-ga.) on 12-in. centers. The joists are easily incorporated into wood framing and are made from cold-formed steel by the same companies that manufacture steel studs, which aren't as deep and have lighter-gauge steel. In use for over four years, the deck has supported up to 50 people at a time with no vibration or discernible deflection. The total design load supported is 77 psf with a $\frac{1}{240}$ deflection.

I have also used light-steel framing to support a clear 20-ft. third-floor span. Steel joists (#80SJ16, which means 8 in. deep by 1¾ in. flange by 16 ga.) on 12-in. centers easily support a third-floor total load of 68 psf with a design deflection of ¹⁄₂₄₀. The 8-in. steel-joist depth saved 3½ in. of headroom compared to 2x12 wood framing.

In commercial work I have supported 140-psf total load with a ¹⁄₂₄₀ deflection using 10-in. deep by 14-ga. steel joists on 12-in. centers over a 20-ft. span; in another case we used 4-in. deep by 14-ga. joists on 6-in. centers over an 11-ft. 4-in. span—all to gain headroom. Gluing with construction adhesive and screwing a plywood structural deck to the steel joists further enhances the stiffness of the composite system.

—*Robert Abramson, Stonington, Conn.*

Chris DeBlois, a structural engineer in Atlanta, Ga., replies: These steel sections are not seen more often in residential floors because of practical reasons, not structural. Structurally, your solution is terrific. In fact, where headroom is an important problem, steel joists may be the only feasible approach. A #925SJ16 has about 27% more strength and almost 90% of the stiffness of a #2 southern yellow pine 2x12. The steel weighs in at only 53 lb. for a 19-ft. span, whereas the 2x12 tops 80 lb. Availability is only a matter of using the Yellow Pages and planning a little lead time (check with your local lumberyard). Prices are comparable: 2x12s go for about $1.35 per ft.; a #925SJ16 costs in the range of $1.55 per ft.

Constructability keeps steel joists from supporting more residential floors. You can't nail steel joists; connections must be screwed or welded. You can't cut them with the same saw you use for wood, and you can't detail them using the standard rules of thumb for wood joists. For instance, in place of solid blocking or bridging, you need to screw steel straps along the bottom flange, spaced according to the stress in the steel. Although the loads for this 19-ft. span are too small to require them, you need web stiffeners when concentrated loads and end reactions get too high. On the other hand, if the crew has experience using steel joists, these issues are minor, and the system can, as you say, be easily incorporated into a wood-frame house.

MYSTERY HOLES

All of the antebellum homes that I have worked on in this area have had 1½-in. dia. holes about 6 in. deep that were drilled in the top of the floor joists and randomly spaced on about 2-ft. to 3-ft.centers. Attic floor joists do not have these holes. The holes seem to weaken the joists considerably, and I do not see that they serve a purpose.

The only idea I have is that maybe an iron rod was placed in the holes and used as a lever to pull flooring up tight. Any other ideas?

—Ben Erickson, Eutaw, Ala.

D. Jameson Gibson, a builder from Charlottesville, Va., replies: I can only offer a theory. I have also come across this configuration in the past and have found the holes to be very useful when re-laying

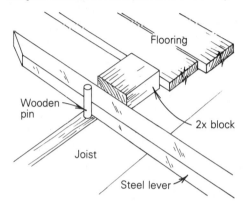

the flooring. I discussed this phenomenon with architect Floyd E. Johnson. The drawing above shows our best guess.

ADDING KNEEWALLS

I'm converting the attic of our small cape into living space. How does one build kneewalls in an attic conversion without transferring roof (and snow) loads to the unsupported ceiling/floor joists below?

—Kris L. Johnson, Madison, Wis.

Peter H. Guimond, a structural engineer from Pawtucket, R. I., replies: Depending on your roof pitch and on the size and span of your existing ceiling joists, there may not be a problem with transferring some of the roof load, but to be sure you would need to have an engineer check out the structure.

The trick to adding kneewalls without transferring roof loads is to connect the kneewall to the rafters without the rafters actually bearing on the kneewall. This way the rafters can deflect under snow loads without affecting the kneewall. On p. 188 is a sketch of a possible solution to the connection problem. It employs a slotted metal connector called a roof-truss clip (Simpson Strong-

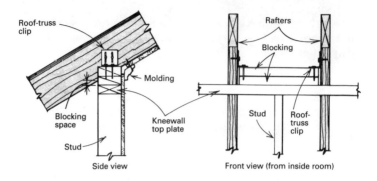

Roof-truss clip

Molding

Blocking space

Kneewall top plate

Stud

Side view

Rafters

Blocking

Stud

Roof-truss clip

Front view (from inside room)

Tie, 1450 Doolittle Dr., P. O. Box 1568, San Leandro, Calif. 94577; 510-562-7775).

Basically the kneewall is built short of the rafters—how short depends on the expected deflection. Blocking is then nailed to the top plate between the rafters. You can place blocking between alternate pairs of rafters to save lumber and labor. And finally, the roof-truss clips are nailed to the blocking and to the rafters. The slots in the clips allow for rafter deflection.

With the framing connection made, the next question is how to keep the drywall joint between the wall and the ceiling from cracking when the rafters deflect. The answer is not to have a drywall joint at that intersection; use molding instead. The molding should be tight against the kneewall but attached only to the rafters. This covers the gap in the drywall and allows the rafters to move.

BUILDING A SHED-ROOF CANOPY

How should I go about building a shed-roof overhang for a 6-ft. atrium door on my home? I've checked many carpentry books and dug through my *Fine Homebuilding* back issues, but I couldn't find anything. Also, if I use wood brackets to support the structure, what is the maximum depth I can go with on the overhang?

—*Fatimah Iolonardi, New York, N. Y.*

Scott McBride, contributing editor of Fine Homebuilding, *replies:*
Building an overhang, or entrance canopy, is fairly straightforward (see drawing on the facing page). The ceiling joists and the rafters can be framed with 2x4s; use 2x6s for the upper and lower ledgers and the subfascia. First nail the lower ledger into the wall studs over the top of the door trim. Mark the locations for the ceiling joists on 16-in. centers and at each end. I think a 30-in. overhang with a 16-in. total rise against the wall works well. Cut the bottoms of the rafters so that the combined height of the ceiling joist

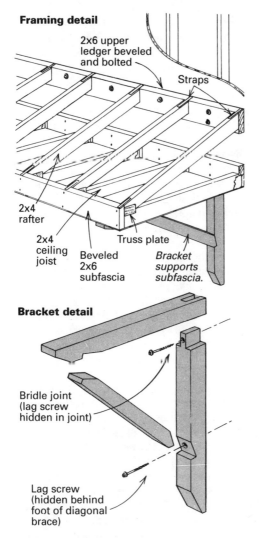

Framing detail

2x6 upper ledger beveled and bolted

Straps

2x4 rafter

2x4 ceiling joist

Beveled 2x6 subfascia

Truss plate

Bracket supports subfascia.

Bracket detail

Bridle joint (lag screw hidden in joint)

Lag screw (hidden behind foot of diagonal brace)

($3\frac{1}{2}$ in.) plus the front edge of the rafter equals $5\frac{1}{2}$ in.—the height of the 2x6 subfascia. Use truss plates to connect the two members.

Bevel the upper ledger and the subfascia to match the pitch of your rafter/joist trusses. Measure the distance of the open end of the ceiling truss to find the position for the upper ledger. Bolt or screw it to the wall and toenail the rafters into it. Use joist hangers to connect the ceiling joists to the lower ledger. Nail on the subfascia, making sure the trusses are square, then add steel straps to beef up the rafter-to-subfascia and rafter-to-upper-ledger connections. Place temporary supports (a couple of 2x4s will do) at each outside corner to hold the entrance canopy level as you install the brackets. Alternatively, you can build the whole thing on the ground and lift it onto the brackets, then bolt it in place.

The brackets are usually constructed of 4x4s. Use good-quality kiln-dried lumber. Construction-grade lumber is likely to check and twist as it dries. This would be particularly objectionable at eye level next to a main entrance.

Tightly nailed or screwed butt joints would probably be adequate to join the bracket components for a small canopy, but some simple joinery will greatly increase the brackets' strength. At the square corner against the building, I would use a bridle joint, where one member is housed in a fork cut in the end of the other member, or at least a half-lap, where both members are notched to half their thickness to make a flush joint. The diagonal brace is traditionally seated in tapered notches at the top and the bottom to keep it from slipping under load. These joints can be easily cut with a bandsaw.

I can't give a set figure for the maximum allowable overhang using brackets, but I can point out some factors to consider. While a post delivers the outboard weight of the roof straight down, a bracket transfers this weight to the wall. Lateral forces are created that push against the wall at the bottom of the bracket and pull away from the wall at the bracket's top square corner. To fasten the brace at the bottom, therefore, you only need a fastener with shear strength. A lag screw or a heavy spike would normally suffice. At the top, however, where the bracket wants to pull away from the house, the fastener will be in withdrawal. Here, you are better off bolting all the way through the wall to get the full strength of the wall itself for support. A long, heavy lag screw would be my second choice after a through bolt. Hide the head of the lower fastener behind the foot of the diagonal brace. The head of the upper bolt or screw could be buried in the tongue of the bridle joint.

The stiffness of the wall is another factor to consider if you're planning a deep overhang. A 2x6 wall handles stress from a canopy better than a 2x4 wall does. A tall wall, such as the front wall of an open two-story stair hall, will be more prone to deformation from the torquing action of a heavy canopy than will a wall that is braced by an intervening second floor.

Finally, the farther down on the wall the diagonal brace reaches, the more it acts like a post. This reduces the push and pull on the wall, and lets you extend the overhang.

SUPPORTING TIMBER-FRAME POSTS

I am building a 1¾-story timber frame and would like to mount the posts in connectors atop a 3-ft. high concrete-block wall. Are the point-load compression capabilities of concrete block adequate for this arrangement? Are there concerns with lateral strengths or stability I should address?

—Jerry Snodgrass, Chico, Calif.

Robert L. Brungraber of Benson Woodworking Co., Alstead Center, N. H., responds: Although you say that the building is a 1¾-story timber frame, without knowing anything more about those stories or the spacing of the posts, there's no way to assess the vertical load in those timber posts. Similarly, there are different kinds of concrete blocks, so I can only offer some generalities in my answer.

Even if the concrete-block manufacturer claims that its blocks can handle the 15,000-lb. point load a post might exert, I wouldn't put point loads on an unreinforced, 8-in. thick concrete-block wall. The post connector is intended to be cast into and tied down to something heavier than the single block it sits in. In your part of the country, you'll need a serious connection. Therefore, for both up and down loads, you should hook the connector onto horizontal rebar within a concrete bond beam placed atop a block wall where every block is filled with grout. You should also place vertical rebar in the blocks; this rebar ties into the rebar in the bond beam and into more rebar embedded in the footing. Using grout with steel reinforcement gives the block greater load-bearing capacity and anchors the post against the vertical accelerations of an earthquake.

If you're going to backfill against that 3-ft. block wall, the vertical rebar tied into the rebar in the footing will prevent the wall from sliding, overturning and breaking.

You'll also have to protect the end grain on the bottom of the posts from direct contact with the steel post connectors. Materials denser than wood can act as ongoing sources of condensed moisture: a situation certain to promote decay. I treat open end grain with a waterborne preservative and/or separate the two materials with a waterproof gasket.

GLUING TIMBER-FRAME JOINTS

I realize that this can get me expelled from the club, but has anyone considered using glue in scarf joints of girts for timber-frame buildings? Or would it be impossible to fit the joints closely enough for a glue joint?

—D. F. Stillwater, Alberta

Ben Brungraber of Benson Woodworking Co. in Alstead Center, N. H., replies: You were well on your way to answering your own questions about glued timber-frame joints. While it is possible to fit the joints accurately enough to glue them, it is very difficult to get them to stay accurately enough fitted. Most timber framers are using wood at a moisture content above its fiber saturation point. As it dries in service, the wood shrinks, the faces distort from their planed flatness, and the timbers may even twist. These are not ideal conditions for any adhesive. Furthermore, the structural adhesives that one might want to use in this application really ought to be clamped tightly and cured with elevated temperatures. These are difficult conditions to create under normal site situations. The so-called "construction adhesives" that we do use on site are elastomerics, good at resisting short-term loads (eliminating squeaky floors) but of little reliable use in resisting long-term loads.

In short, gluing heavy solid-sawn timber joinery is not worth the trouble of doing it right, nor the mess of doing it incompletely. To answer the rest of your letter: yes, many people have considered using glue in timber joints (sadly, quite a few of them local building officials).

GUNSTOCK POSTS AND BRACES

I'm designing a timber-frame barn that will have gunstock posts, and I'm having difficulty with the proportions of those braces that lie in the same plane as the gunstock swell. To my eye, option A (drawing, facing page) looks better than option B. Can you suggest some guidelines?

—William Stilwell, Alberta

Tedd Benson of Benson Woodworking Co., in Alstead Center, N. H., replies: I'm assuming the guidelines you are looking for are structural rather than aesthetic. In any event, in the construction of a barn (or just about anything) you should satisfy the former before addressing the latter.

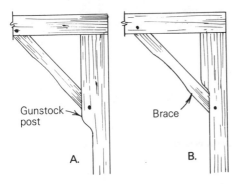

Gunstock post   Brace

A.   B.

The reason for a gunstock configuration on a post is to increase the amount of wood available for multiple joint intersections. In old buildings, the enlarged gunstock area is very often the base of the tree from which the post was made; the post that emerged after hewing was a clear reflection of the shape of the log. Today, with timbers milled as rectangular sections, a gunstock involves significant reshaping of the timber. Therefore, when shaping a gunstock, try to follow the natural direction of the grain to prevent the gunstock taper from directly exposing end-grain. The straighter the grain, the more gradual the taper should be.

For support of the beams, the enlarged part of the gunstock ought to include at least the upper quarter of the post. It is almost always concluded in the upper half. Your barn frame probably gives you some latitude regarding brace length, but very often the proper diagonal length of the brace is about one half the height of the post from floor to beam. When you put these two considerations together it becomes obvious that the gunstock taper and the brace want to collide. In general your option A looks better structurally and aesthetically. I will add, however, that there ought to be at least 6 in. of wood between the bottom of the brace and the beginning of the gunstock taper. There is certainly no problem with your option B, but it would look better with a long brace and a short gunstock so the lines of the transition can be appreciated.

LAYING OUT A DOMED ELLIPSE

I want to build an oval-shaped room topped with a domed ceiling framed with semicircular arches. The exterior of the room would be rectangular, and the interior walls would be framed in an oval shape. How do I lay out such a structure?

—Ray Wernersbach, Milltown, N. J.

Contributing editor Scott McBride replies: To get a handle on the geometry of this problem, hard-boil an egg. Peel the egg, slice it in half along its long axis, and lay a half-egg on a level surface with its flat side down. You're now looking at a model of your domed elliptical ceiling—a little pointy at one end, perhaps, but close enough.

Now slice the half-egg perpendicular to the long axis at ¼-in. intervals. Separate the slices; you'll see that they're more or less semicircular in shape. The diameter varies considerably, with the smallest diameters at the ends of the egg and the largest in the middle. The curved edge of each slice represents a semicircular arch, a series of which you must build to frame your dome.

Before you can tackle the roof, however, you have to frame your elliptical walls. To lay out the ellipse on the floor of a rectangular room, begin by snapping centerlines perpendicular to the long and short walls of the room. These will become the major and minor axes of the ellipse (see drawing on the facing page). Mark the axes one stud width in from each existing wall. These marks will serve as the ends of each axis. Now swing an arc, using one end of the minor axis as the center point and using half the length of the major axis as the radius. We'll call the center point A. The points where the the arc cuts the major axis—the "foci" of the ellipse—will be B and C.

Now drive nails at points A, B, and C, and stretch a loop of string tight between the three points, tying the ends of the string in a knot. Put a pencil point inside the loop at A and remove the nail. Swing the pencil around the room, keeping tension on the string, and draw a line on the floor. The resulting curve is the inside framing line of your elliptical wall.

Next, draw a series of parallel lines on the floor at 16-in. intervals, perpendicular to the major axis. These lines represent the plan projection of the ceiling arches you'll be building. Make your center arch first and then work outward in matched pairs. Using the radii taken from the lines on the floor, lay out a series of arches on any suitably large surface, using a beam compass: a length of wood with a nail in one end and a pencil on the other. The radius of each arch will be half its diameter (half its length as seen in plan).

You can build the arches using any number of methods, but the easiest would be to "bricklay" overlapping pieces of plywood or lumber (see *FHB* #67, p. 12). Fortunately, you have a different radius and span for each pair of arches, which means that you can use up much of the waste that normally occurs when cutting curved plywood members. All you have to do is cut the long sections for the broadest arches from the middle of the plywood sheet, and save the edges of the sheet for the successively tighter curves. Your end cuts will be along the diameter of the circle.

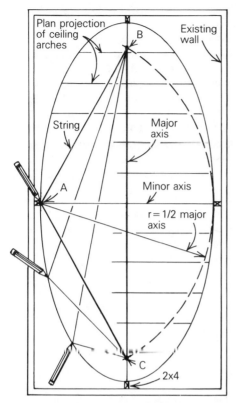

Plan projection of ceiling arches
B
Existing wall
String
Major axis
A
Minor axis
r = 1/2 major axis
C
2x4

The number of layers and their depth will depend on whether or not the ceiling will be self-supporting. If you can suspend the ceiling from an overhead floor or roof of adequate strength, then your arches could consist of two layers of ¾-in. plywood about 6 in. wide. Fasten them together with long drywall screws or clinched nails (to clinch a nail, drive it through both layers and bend the end over). For more strength, sandwich a layer of lumber between the outer layers of plywood (this makes a good nail base for ceiling lath as well). If your dome will be free-standing and will bear the weight of a roof, I suggest that you consult an engineer before fabricating your arches. He may suggest gluelam beams or some kind of curved steel trusses.

Install the arches by tipping them up on the walls like ordinary trusses, bracing them as you go to keep them plumb. Finish the inside of the dome with plaster over wire lath. To reduce weight, apply the scratch and brown coats using a gypsum plaster with a perlite aggregate.

If you got an early start it'll be lunch time by now and you'll be good and hungry. Go to the kitchen, open a cold bottle of soda and whip up some egg salad.

CURVED WALL WITH BEVEL SIDING

In the process of renovating a cupola dome on a historic-landmark courthouse, a carpenter friend needs to apply 6-in. redwood bevel siding to a round 22-ft. dia. dome base. A problem arises because the siding must bend around a smaller circumference at the top of the board and a larger circumference at the bottom where it overlaps the previous course. Even after a great amount of steaming, the siding tends to bend upward and buckle out at the top. Is there a method for making this application with bevel siding?

—Joe Russo, Washborn, Wis.

Architectural woodworker John Leeke of Sanford, Maine, replies:
The easiest way to handle this problem is to use bevel siding that's rabbeted on the bottom edge. You can make this yourself, but it's also available commercially from some manufacturers (drawing below). Both edges wrap around the same circumference, the back lies flat against the wall and the board lies level with no bow up. Some steaming may still be necessary to get the butt ends to lie flat.

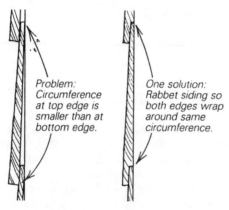

Problem:
Circumference
at top edge is
smaller than at
bottom edge.

One solution:
Rabbet siding so
both edges wrap
around same
circumference.

The only problem with this method is that the the siding's lower edge will be thinner by half. When sunlight falls on the finished wall the smaller shadow cast by the thin edge will give the wall a distinctly different look than full-thickness siding would. If this is unacceptable, then your friend needs to think like a boatbuilder to solve this problem.

Because of the overlap of the siding, the lower edge of the bent board has to be longer than the top edge. Your friend needs beveled siding boards that are curved. He can make them himself with a bandsaw and a thickness planer.

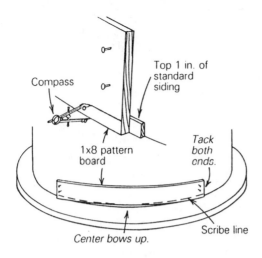

Compass

Top 1 in. of standard siding

1x8 pattern board

Tack both ends.

Center bows up.

Scribe line

The first step is to make a pattern board. For this, use a standard 1x that has no bevel yet and that's a bit wider than the siding (a 1x8 in your friend's case since his siding is 1x6). Next, rip the amount of overlap, usually 1 in., off the top edge of a piece of standard siding. This represents the preceding course of siding and is used to shim out the bottom edge of the pattern board. Tack this piece to the back of the pattern board, as shown in the drawing above.

If the wall that's being sided has a level base or water table, then bend the pattern board around the wall and nail the ends flat so that the bottom edge touches the base. When this is done, the center of the board will bow up above the base.

Next, set a pencil compass to the maximum distance between the base and the bottom edge of the pattern board. Now, moving one side of the compass along the base, scribe a line on the pattern board, as shown above. This scribed line represents the amount of curve you need. Lay out the top edge parallel to the scribed line. Then remove the pattern board from the wall, cut out both lines on the bandsaw and you have the pattern.

Since your friend is working on a cupola, he probably doesn't have a level base to scribe along. In this case, he needs to trace a level line onto the wall with a 4-ft. level and then nail a strip of wood on the line, anything for the compass to ride against. Now, he can proceed as described above.

To make the siding, start with boards that are the thickness of the lower edge of the siding and wide enough to accommodate the curve. Use the pattern to mark them, and then cut out the curved boards.

The last step is to bevel the bowed boards with a thickness planer. Build an auxiliary table out of plywood for the bed of a planer. It should be canted to hold the boards at the correct angle to the cutterhead for the bevel to be cut. Add a low fence on both sides of the new bed. Make them curved to match the edges of the bowed boards. Now plane on the bevel.

Of course, you should test the first few boards by installing them on the wall, just in case your pattern needs some adjustment.

INSIDE CORNERS FOR BEVELED SIDING

We are renovating an old house (circa 1900) with 6-in. beveled cedar clapboard siding. We have run into a major snag: how do you cut beveled siding so there is a clean line of siding at the inside corners, especially at angles (45° or 60°), without using vertical wooden or metal strips to cover the seams? Lots of old houses around here have this detail, but without taking them apart, we are stumped about how to do it.

—Linda Canton, Minneota, Minn.

Consulting editor Bob Syvanen replies: I usually use a small vertical corner stick (¾ in. by ¾ in.) at inside corners when I'm installing beveled clapboards, and simply butt the clapboards into it. But I

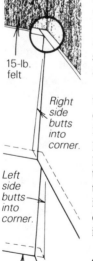

15-lb. felt

Right side butts into corner.

Left side butts into corner.

6-in. clapboards

have installed shingles without corner sticks, weaving the inside corners. Clapboards could be done the same way, with a lapped joint, alternating courses. In either case, I always start with a strip of 15-lb. felt in the corner as a precaution.

After nailing up the starting strip, which shims the first clapboard to the correct angle, the first course, on the left side for instance, is cut to butt into the corner, as shown at left. The right-side piece is cut on an angle to fit against the first piece. On the second course, the clapboard on the right side goes up first and is cut on an angle to butt into the corner. The piece on the left side is cut to fit against the piece on the right, lapping over it, and so on. The result is a tight, clean inside corner.

If the corner isn't 90°, the procedure is the same, but the cut on the end of the clapboard will be a compound angle. Set the bevel of your saw to the angle of the corner.

**I will soon be installing some cedar shingles on a portion of
a house I designed. As shown below, I am installing the
shingles radially at the top of an arch. Trimming each**

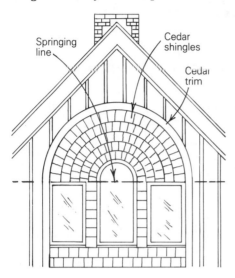

Springing
line

Cedar
shingles

Cedar
trim

**shingle to a wedge shape is no problem. My question is what
happens toward the springing line where good practice
would dictate both an endlap and a sidelap condition?**

—*Gregory S. Kindig, Harrington, Del.*

Dave Skilton, a builder in Coburg, Ore., replies: As you have already
realized, your design exceeds the limitations of shingle siding,
which is essentially a horizontal lap system. Even a rather elaborate
solution, like the one proposed here, won't guarantee a completely
watertight condition, especially if your wall gets wind-blown rain.
Some other form of waterproofing behind the shingles is indicated.
Once you fall back on that secondary system, the shingles have
become purely decorative, and you can do what you want. To me,
however, the beauty of decorative shingles lies precisely in the fact
that they are also functional.

With that in mind, I offer the following idea. Because the shingles
will have to be specially sawn along their length anyway, why not
set the sawblade at an angle so that the shingles are beveled and

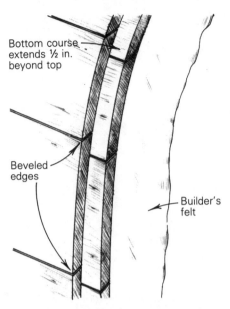

Bottom course extends ½ in. beyond top

Beveled edges

Builder's felt

overlap at least a little along their sides (drawing above). I also think it might be wise to double up the courses, at least in the arch section, using lower-grade shingles underneath to keep the price from doubling. You could even get a little extra shadow drama by slipping this bottom course down ½ in. Solid cedar trim boards between the windows might help make the transition in the pattern from the horizontally installed shingles at the springing line to the vertical ones below. (For those who are wondering, the springing line is the imaginary horizontal line at which an arch begins.)

FIRE-TREATING CEDAR SHAKES

I have a low-cost source for heavy butt #1 red-cedar shakes. Is there some way I can treat the shakes with fire retardant before I apply them? If so, what retardant should I use, and where can I get it?

—Randall J. Gerlach, Brookings, Ore.

Michael M. Westfall, president of the Cedar Shake and Shingle Bureau, replies: The only permanent fire-treating method that is recognized and approved by building codes (UBC standard 32-7) is the full-cell vacuum-pressure process. Shakes and shingles are locked inside a chamber where a vacuum is created, and fire-retardant chemicals are injected into the innermost layers of the cedar. Several treating companies have been authorized by the Cedar Shake and Shingle Bureau to perform this process. In your area, Cascade Pacific Industries (Jasper Treating Div., P. O. Box 1351, Eugene, Ore. 97440; 800-331-0656) treats shakes using the approved process.

If you're bent on doing it yourself, the Texas Forest Products Laboratory (P. O. Box 310, Lufkin, Tex. 75901) has found that liberal applications of diammonium phosphate (a chemical fertilizer) dissolved in water achieve the UBC's minimum fire-resistance rating. Unlike the full-cell vacuum-pressure process, this coating is not permanent. If the roof is in a fire-prone area, the researchers recommend monthly applications, especially during the fire season.

BOARD-AND-BATTEN SIDING

We are going to re-side an old barn with board-and-batten siding, using poplar 1x12s for the boards. What is the correct way to face the boards with regard to their annual rings? Also, what is the correct nail pattern for board-and-batten siding?

—James Blendening, Wallingford, Pa.

Senior editor Charles Miller replies: I checked with the California Redwood Association and with the University of California's Wood Building Research Center. The bark side of the board should be oriented toward the exterior (for any species of wood). That means the convex side of the annual rings should face out.

As for nailing, pairs of nails should be used to affix the 1x12s to the wall. They should be on 4-in. centers, dividing the board into thirds. Ring-shanked nails or hot-dipped galvanized will work fine, but I recently learned of someone who swears by galvanized drywall screws. The 1x12 boards should be spaced ½ in. apart. The battens should be attached with a single nail (or screw) that passes through the ½-in. space. According to the California Redwood Association, the boards should be nailed to horizontal 2x blocking installed between studs at 24 in. o. c.

RE-SIDING AN OLD HOUSE

We recently purchased a 100-year-old Victorian house in Illinois. The wood siding was badly neglected, and we're considering putting up vinyl siding. Should we retrofit some type of vapor barrier before installing the vinyl?
—Theodore R. Knack, Johnson City, N. Y.

George Nash, author of Renovating Old Houses, *replies:* I don't recommend vinyl for re-siding old houses. Placing vinyl over the old wood siding can mask or even cause water damage. Also, vinyl siding can act like an exterior vapor barrier, trapping moisture within the walls and abetting decay. In very cold weather, vinyl cracks easily, and its color will fade over time. Instead, remove the old siding and consider using fiber-cement siding (FibreCem Corp., P. O. Box 411368, Charlotte, N. C. 28241; 704-588-6693) as an alternative to wood. This new product, warranted for 50 years, is a composite of portland cement and cellulose. Fiber-cement siding is installed like any lap siding, but I'd use stainless-steel nails. It's best to nail directly into studs unless you've got beefy sheathing.

Fiber-cement siding comes in 12-ft. lengths, and you butt the lengths with joiner clips that automatically align the two pieces. For big jobs you should invest in a diamond-tip masonry blade to cut the siding, but you could get away with a masonry cutoff blade on a smaller job. Although you don't have to paint fiber-cement siding (its color is very close to that of concrete), I would imagine a Victorian house would look better painted. The nice thing about fiber-cement siding is that once it's painted, the paint job lasts about 10 years. Fiber-cement siding is noncombustible; it doesn't rot, dent or crack, and termites won't eat it. The siding comes in different textures, including wood grain, and it costs about the same as vinyl siding.

Whatever siding you do use, it's absolutely critical that you leave the original trim intact. Although stripping and repainting are undeniably labor intensive, to yield to the temptation to cover the trim with vinyl siding would do irreparable harm to the aesthetic and architectural character of the house.

When you put up the new siding, make sure the seams between it and the old trim are tight. Cut rabbets in wood drip caps and sills to channel water away from the siding and use copper or galvanized-steel flashing (both of which are superior to aluminum) to prevent possible water traps. I often use prepainted galvanized coil stock (Ideal Roofing Co., Ltd., 1418 Michael St., Ottawa, Ont., Canada K1B 3R2; 613-746-3206). Don't rely solely on caulking; even the highest-quality acrylic copolymer caulking, such as Geocel

(Geocel Corp., P. O. Box 398, Elkhart, Ind. 46515; 219-264-0645) will eventually fail if it's the only defense against water penetration.

As far as tightening up the house itself, apply an anti-infiltration membrane, such as Tyvek, over the sheathing before installing the new siding. This membrane prevents outside air from getting into the house, making it easier to maintain internal humidity while reducing the chances of condensation. Wherever possible, slide the membrane up under the trimboards. With the old siding removed, you can also blow additional fiberglass or cellulose insulation into the wall cavities before re siding.

Although it's not mentioned in your question, I'm assuming the interior walls are plastered. Old three-coat plasterwork is actually a fairly good vapor barrier, especially when painted with a vapor-retardant paint. Caulk all the joints between trim and wall surfaces before applying the paint.

Adding storm windows or double-glazed replacement windows is a necessary part of any moisture-control program, and so is the installation of a vapor barrier over a bare-earth cellar floor. This vapor barrier should be a heavy polyethylene that's covered with an inch of sand and 3 in. or 4 in. of crushed stone. Alternatively, you could pour a concrete slab over the poly. All of these steps will go a long way toward curing any moisture problems.

AIR/VAPOR BARRIERS AND VINYL SIDING

My new home is finished with exterior vinyl siding, manufactured by Certainteed Corp., with a 25-year warranty. I question the fact that the builder didn't install 15-lb. felt or other type of vapor barrier over the sheathing before installing the vinyl siding. Would this be the normal construction method for installation of vinyl, and if not, would this void the warranty of the siding?
—Thomas Campo, Wappinger Falls, N. Y.

Associate editor Kevin Ireton replies: Air/vapor barriers, like builder's felt or housewraps, are applied over sheathing (and under siding) primarily to keep the wind from whistling through your walls. They're a good idea, but certainly aren't mandatory. The lack of an air/vapor barrier should have no effect on the siding. I called the folks at Certainteed and was told that this will not void the warranty on your siding.

PROBLEMS WITH A GRAVEL ROOF

About two years ago, I bought a contemporary Ranch-style home, built in the mid-60s. The house has a low-pitched roof with a rise of 1 ft. for every 8 ft. of run. The current roofing is hot tar covered with pea gravel and was redone three years ago. How do I maintain this type of roof? It is very messy to take care of. I have had to clean the gravel out of the gutters twice a year, removing nearly 200 lb. of gravel each time. The gravel also falls onto our sidewalks and patios with each rain. Also, are there any alternatives to this type of roofing when it needs "resurfacing" the next time?

—Gordon R. Huey, Hope, Ind.

Harwood Loomis, a consulting architect in Woodbridge, Conn., replies: Your roof pitch, 1½-in-12, is actually toward the steep end of the range for low-slope roofs. With this slope, the asphalt used under the hot tar should have been either Type III (Steep) or Type IV (Special Steep). Because you are losing gravel, I'd have to guess that the roofer used Type II asphalt, which has a lower softening temperature and thus could allow the gravel to migrate downslope when the hot weather arrives. Or he might have put on too much gravel or waited too long to apply it. The gravel provides some ultraviolet (UV) protection for the asphalt and the membrane, and it is supposed to be embedded in the hot asphalt so that it will stay in place.

Although some gravel loss is normal, the amount you're losing seems excessive. Gravel should be applied at a rate of 500 lb. per square (100 sq. ft.) or 5 lb. per sq. ft. (+/- 25%). If you are collecting 200 lb. each time you clean up, that's how much gravel should be on 40 sq. ft. of roof. I would suggest that you go up on the roof and sweep a sample section with a stiff, shop-type broom. See how much gravel is actually embedded in the asphalt. If you can sweep the loose gravel off the surface and still leave a continuous coating of gravel, then there is just too much up there, and you can safely sweep off the excess. If moderate sweeping leaves nearly no gravel, the asphalt was too cold when the gravel was applied.

In the latter case, you will have to make a decision about what to do. You could have a roofer sweep away the gravel, flood the roof with a thin coat of hot asphalt and replace the gravel. The other option would be just to leave things alone and to clean up periodically whatever gravel finds its way into the gutters or off the roof.

Maintenance for an asphalt, built-up roof consists mainly of an annual inspection and keeping drains open. If you find blisters in the roofing in hot weather, it's an indication of moisture in the membrane and should be remedied by a roofer. Small blisters can remain if they are isolated. But if the blisters are large or numerous, the roof may no longer be firmly attached to the deck. In such a situation, the blisters should be cut open, dried and then patched with roofing cement. This is not something I recommend home owners try themselves, however.

At some point, either blistering will become excessive or there will be considerable bitumen bleed-out and erosion of the membrane. That's the point where you will need to reroof. There are processes available that claim to "resaturate" the old, dried-out membrane. However, I consider these products to be snake oil; they don't work. You can get a few more years of life out of a roof, if the deck and the insulation are dry, by putting a new two-ply, built-up membrane over the existing roof, but I believe the best approach is to bite the bullet, remove the old roof and install a new one.

In lieu of the mess of a built-up roof, you have the option of using any of several single-ply or modified bitumen roof membranes. If your roof is visible from the ground or from inside the house, you might be most pleased with a granule-surfaced, modified-bitumen membrane. The granules provide UV protection, and you can select a membrane with granules that complement the color of the building (for more information on modified-bitumen roofs, see *FHB* #64, pp. 43-47).

COLD-ROOF DETAILS

I have questions about two roof details I've seen in the mountain area of central Idaho. The first is a cold roof. From the top of the roof down it consists of #1 cedar shakes, 30-lb. roofing felt, ¾-in. plywood roof sheathing, 2x4 sleepers laid on edge at 24 in. o. c., 1x6 skip sheathing at 6 in. o. c., a Tyvek air-infiltration barrier, 2x12 rafters at 24 in. o. c., R-38 fiberglass-batt insulation, a 4-mil vapor barrier and the finish ceiling. My inclination, however, would be to reverse the skip sheathing and the plywood sheathing and to replace the Tyvek with 30-lb. roofing felt installed over the plywood. This would let the cedar shakes breathe and would provide a waterproof layer to shed condensation (the detail as it stands won't stop condensation from entering the house). Applying plywood instead of skip sheathing directly to the rafters would also increase the roof's shear resistance.

The other detail consists of 26-ga. metal roofing on 1x6 skip sheathing 8 in. o. c. over Tyvek over 2x10 rafters. I'm afraid that condensation from the metal roofing may drip right through the Tyvek and end up in the insulation and finish ceiling.

—David Kalange, McCall, Idaho

Steve Kearns, a builder from Sun Valley, Idaho, replies: Your concerns are justified. To provide the best shear diaphragm with the closest connection to the walls, plywood sheathing is normally installed over the rafters rather than over the sleepers (drawing below). Skip sheathing is recommended on top of sleepers where wood shakes are used. This lets shakes breathe, especially in areas of high humidity (here in Sun Valley we have very low humidity, so it's not nearly as important as in a high-humidity climate). We use Simpson A-35 framing clips for connecting the sleepers to the plywood. We use a nailing schedule developed by a local structural engineer.

You're also correct in assuming that the Tyvek should be replaced with 30-lb. felt installed over the plywood. You do need a waterproof barrier on top of the plywood to catch condensation, and some condensation will occur on even the best cold roof. Some architects in our area have gone so far as to specify double layers of 30-lb. felt over the plywood with the sleepers set in mastic. While I think this is overkill, it's all designed to stop condensation

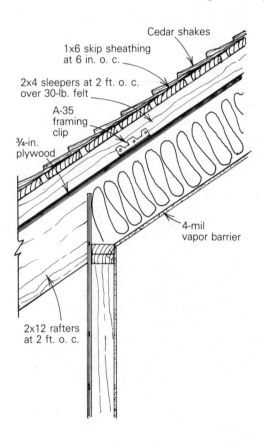

Cedar shakes

1x6 skip sheathing
at 6 in. o. c.

2x4 sleepers at 2 ft. o. c.
over 30-lb. felt

A-35
framing
clip

¾-in.
plywood

4-mil
vapor barrier

2x12 rafters
at 2 ft. o. c.

from leaking through the plywood and into the insulation and the ceiling below.

The metal-roof detail you mention is a poor one. My roofing contractor strongly recommends that metal roofing be placed on solid sheathing. As you point out, condensation occurring below the metal needs to be caught by a waterproof barrier. Tyvek or roofing felt over skip sheathing will tend to "pocket" between the sheathing boards, encouraging puddles of condensation to form— and eventually leading to a leak. You could place metal roofing over skip sheathing on a cold roof (and a cold roof makes sense even for metal roofing, especially where there are valleys, or where there's not enough pitch to ensure that snow will always slide off). But even on a cold roof, solid sheathing eases the installation of a metal roofing: you can predrill all your screw holes without having to worry about the skip spacing, and you need not worry about getting a heavy enough gauge of metal for your skip span. Lighter gauges of metal are cheaper, too.

ICE-DAM MYTHS

I have an ice-damming problem. My roof consists of cedar shingles over skip sheathing. The roof has a 7-in-12 pitch with a 2-ft. overhang, a dripedge vent runs along the top of the rafter-tail plumb cut. The attic insulation is R-40 with properly installed baffles. There is about 8 sq. ft. of attic venting (gable-end louvers and roof vents) for 1,400 sq. ft. of attic space. The lower 36 in. of the roof is covered with Bituthene (W. R. Grace Company, 62 Whittemore Ave., Cambridge, Mass. 02140; 800-242-4476), but because of the overhang I get only about 4 in. of coverage past the top plate. To combat the problem, I want to add a second layer of Bituthene that extends farther up the roof and overlaps the first layer by 6 in. Will this solve the problem? Can you offer any other advice?

—Jeffrey H. Goldsmith, Bolton, Mass.

Gene Leger of Leger Designs in New Boston, N. H., replies: Bituthene won't stop ice dams; that's not its function. The membrane will provide some protection in case of water backup, but it won't prevent ice from damming or water from backing up the roof and under the shingles in the first place. Ice damming is a sign that there is too little insulation, not enough ventilation or both.

There are three widespread but mistaken beliefs at work here. The first is that Bituthene stops ice dams. The second is that combining different types of vents will provide more ventilation.

The third mistaken belief is that well-insulated and well-ventilated attics are always free of ice dams.

Your drip-edge vent, gable-end vents and roof vents notwithstanding, my guess is that you have inadequate ventilation. One problem is that the total net free venting area of your roof vents is too little to balance that of your soffit vents. Another problem is that the three different types of vents are fighting each other. Gable vents work only when the wind is blowing directly at them—that is, over the insulation. Soffit vents, on the other hand, don't rely on wind direction, but their effectiveness is reduced by air flow over the insulation.

I suggest that you remove the existing roof vents and install a continuous ridge vent with an external baffle (the baffle prevents wind-driven snow and rain from entering the vent). Remove the drip-edge vent and install a double louvered soffit vent next to the fascia board. Then block up the gable-end vents. By blocking them on the inside, rather than removing them, you eliminate the problem of having to patch the siding.

READERS REPLY

I read Gene Leger's solution to Jeffrey Goldsmith's ice-dam problem on his new home (see p. 207) and suspected that Leger had not heard the whole story before he made his recommendations. Goldsmith's R-40 attic insulation level sounded fine, and the roof venting, while not optimal, didn't seem so far off as to cause the trouble he was experiencing. I suspected that there was another factor involved.

Ice dams are caused when the snow on the roof begins to melt. Water then runs down the roof and freezes, usually at the roof overhang. The heat that melts the snow comes from inside the attic. Ice dams are usually blamed on inadequate ceiling insulation and/or inadequate roof ventilation; the explanation being that the heat conducted into the attic isn't removed quickly enough by the ventilation.

More recent investigations by Gary Nelson of the Energy Conservatory in Minneapolis, Minnesota, have shown that in many cases ice dams are caused by uncontrolled air leakage into the attic, called attic bypasses. Typical attic bypasses include bath fans, recessed lights, chases around chimneys, plumbing and electrical chases, attic hatches and interior partition walls (especially in balloon framing). Any hole in the ceiling below the attic can act as a little chimney, funneling warm air from inside the home to the underside of the roof sheathing, which warms up and melts the snow above. In addition to attic bypasses, any source of heat in the attic that raises the temperature inside can cause an ice dam.

My curiosity got the best of me, so I called Goldsmith. I learned that he had forced-air heat, and the heating ducts for the second floor are located in the attic. Bingo!

The registers for the second floor are located in the ceiling adjacent to the exterior wall. Insulated flex duct is used in the attic. In addition, because the second floor appeared to need little heat, and the system is one zone, Goldsmith had closed all the second-floor registers. I surmise that the flex-duct connections at the register boots were leaking (almost all ductwork joints leak) and dumping hot air right at the eave all around the attic. The closed registers exacerbated the problem by raising the pressure at the leaky joints. You almost couldn't invent a better system for making ice dams!

The solution includes doing the following: make sure that all ductwork joints in the attic are carefully sealed; install adequate insulation on the register boots themselves (spray foam may be the best solution); seal the register boots to the ceiling drywall so that house air doesn't leak into the attic; and increase the insulation on the ducts themselves. If the second floor needs to be zoned separately from the main level, tight-fitting dampers should be installed in the ductwork in the basement. All these changes will lower the energy use of the house, too, because the heat will end up in the house instead of in the attic.

In closing, I should add that putting heating ducts in any unheated space in a heating climate should be avoided whenever possible.

—Marc Rosenbaum, P.E., of Energysmiths, Meriden, N. H.

SLATE-ROOFING PREPARATION

I am about to begin an extensive renovation project that will take seven or eight months to complete, including the installation of a slate roof. I do not want the long lead time for the delivery of the slate (which may easily be several months) and the long installation time to interfere with our schedule. Can you recommend a temporary roof that will weatherproof the structure, and that can be left in place, with the slate installed over it?

Additionally, this project is located on the shore and is subject to substantial wind during storms. I would like to know if you have any tips to increase the weather-tightness of a slate roof where wind-driven rain is common. I am particularly concerned about the gable ends.

—Paul Hirsch, Stamford, Conn.

Todd Smith of Rogers Roofing Co. in Verona, N. J. replies: The long lead time is a common problem with slate roofing. So to keep the construction project on schedule, we've developed the following procedure for keeping the building dry while we wait for a slate delivery. First we fabricate and install all of the flashing details, such as the crickets, step flashings, counter flashings, soil-pipe flashings and valleys. When we install the flashings we use short roofing nails (1 in. or less). This will allow us to remove them and adjust the flashings as required when the slate is installed.

In conjunction with the flashing installations we install the roofing felt. We always use 30-lb. roofing felt on a slate roof. If the felt is to be exposed to the weather for a long time, or if the roof is not very steep, we install the felt with an 18-in. exposure. When installing the felt, we use roofing nails rather than staples because the nails will hold better. We also nail pieces of lath over the edges of the felt to keep the wind from getting underneath and blowing the felt off.

To protect the felt and to make the roof worry-free, we nail lightweight poly tarps on the roof. We use the smallest tarps possible so no part of the tarp hangs over the roof to catch the wind. We nail the edges of the tarps down through lath strips. If necessary, we nail some lath strips in the center of the tarps for more support. We use short nails that do not go clear through the sheathing. This temporary roof will remain water-tight for months, adding little cost to the job.

To increase the weather-tightness of a slate-roof installation there are several procedures you can use. The first I've already mentioned, which is to install the roofing felt with only an 18-in. exposure. The second procedure is to decrease the amount of slate exposed to the weather, which will increase the headlap. Wind-driven rain will then have to blow further up and under the slate to leak.

The third procedure is to install a course of 30-lb. felt in between each row of slate. The felt should be cut to the same length as the slate. For instance, if the slates are 18 in. top to bottom, the felt should be cut 18 in. wide. Install the felt in the longest lengths possible and nail it only at the top. The felt will help prevent water from getting under the slate. To keep the slates on the gable end from leaking we set them in cement, almost the way you'd lay brick.

ROOF ASSEMBLIES FOR HOT CLIMATES

I'm designing a home in Texas that will have great emphasis on energy conservation. However, I've seen very few cathedral-ceiling roof designs for hot climates. Would an adaptation of the cold-roof concept work? Three inches of air space above the sheathing would allow venting of heated air, which often exceeds 130° F. The 2½-in. air space between the sheathing and the fiberglass ceiling insulation would vent whatever heat does penetrate the assembly. Is this plan feasible?

—Robert Howell, Arlington, Tex.

Larry Maxwell, senior research architect at the Florida Solar Energy Center, replies: Although insulation is important, the first goal in a hot climate should be to keep radiant heat out of the roof assembly in the first place and to find a way to expel any heat that does get in. This requires that you reduce the heat energy source, control the transfer of heat to the interior, and properly ventilate the roof assembly. A very effective roof consists of (from the underside up) drywall, a continuous vapor barrier, insulation, a minimum 2-in. airspace, a radiant barrier, roof sheathing, roof felt, and either shingles, metal roofing, or roofing tiles (drawing below). This assembly works for both hot-and-humid or hot-and-dry climates.

The energy source is the sun, acting through the finish roofing. You can reduce its effects by carefully choosing the material and color of your roofing. Shingles should be as lightly colored as possible. But even white shingles can absorb a lot of heat, as they

A hot-climate roof

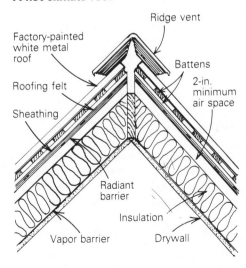

do not have a very high reflective finish. If you use shingles, coat them with a white elastomeric or acrylic roof paint. The paint will make them more reflective and lower their surface temperature by 30° F to 40° F. Two better alternatives to shingles are metal and tile. The surface temperature of a metal roof with a factory-painted white finish will remain close to the ambient air temperature on even the sunniest days. You can create additional ventilation by installing the roof over two layers of battens: the first layer running up the slope (to create a drainage route for trapped moisture) and the second layer running perpendicular to the first. A tile roof is installed in a similar manner. An airspace is created by a combination of battens and the shape of the tile itself. But tile has an additional advantage. The mass effect of concrete or clay will slow down the transfer of heat to the roof deck, regardless of the roof color. And the very high emissivity of these materials means that they'll re-radiate stored heat at night, actually cooling the roof.

To reduce the transfer of any nonreflected radiant heat to the interior, install a radiant barrier. It should be placed between the roof sheathing and the rafters, with the shiny side facing down. It can be draped over the roof rafters before the sheathing goes on, or attached to the back of the roof sheathing before installation. The second method is less labor-intensive than the first, but requires a perforated radiant barrier. You must leave a minimum 2-in. airspace between the radiant barrier and the insulation below it.

To ventilate the roof, I recommend continuous soffit and ridge vents. A manufactured ridge vent is preferable to a site-built one. In the ceiling, 9 in. of insulation should be adequate. Install the insulation neatly and pay special attention to the area at the ridge beam; poorly installed insulation can fall away from the ridge, leaving an uninsulated space. It's also important to keep the insulation from blocking the soffit vents. When installing your vapor barrier, make sure that it isn't compromised by electrical or piping penetrations, or by such items as recessed light fixtures.

METAL ROOFING ON A FLAT ROOF

My house has a built-up roof that is flat on one part and low-pitched on another part. I would like to cover the entire roof with metal roofing to prolong its life and add further protection against leakage. However, I've heard that it's bad design to install metal roofing on a flat roof. Is this true? If so, what are the reasons, and can they be circumvented?

—William Bernard, Los Alamos, N. Mex.

Matt Holmstrom, a remodeling contractor who specializes in metal roofing, replies: I do consider a flat metal roof to be poor design. While it is possible to install a locked-seam roof on a very low slope, this is a particularly troublesome and expensive installation. Briefly, it consists of small (approx. 18 in. by 24 in.) rectangles of metal, flat-seamed and soldered together on all four sides. Not only is this much soldering extremely difficult and time-consuming, but later expansion and contraction can break open some of the joints and ruin what appeared to be a good installation. Also, if terne or galvanized steel is used on such a roof instead of more expensive copper or stainless steel, any standing water will peel the paint and greatly accelerate rusting.

I think that any residential roof that doesn't shed water is a poor design, though I know that flat roofs are important to the architecture of your region. My first solution would be to reframe the roof with a pitch that is adequate for standing-seam metal (at least 1½-in-12). If this can't be done without violating the appearance and architectural integrity of your home, I recommend using a membrane roof instead of a built-up or metal roof. I've seen good results over the short term from modified bitumen. While I can't vouch for it over the long haul yet, it costs about half the price of a terne roof.

ALUMINUM SNOW GUARDS

Our home in Vermont has a two-story gable roof with a shed-roof carport attached at the eave. Both roofs are aluminum. Snow and ice build up on the upper roof and eventually melt enough to come crashing down on to the carport, deforming and even puncturing the aluminum roof.

I want to install snow guards on the upper roof to avoid this problem, but I can't find cast-aluminum snow guards. I can find galvanized snow guards, but I'm worried about galvanic reaction. Is my fear real? If so how could I avoid it?

—Ross S. Gardner, Livingston, N. J.

Editor Kevin Ireton replies: The potential for galvanic reaction between aluminum and galvanized steel is pretty small. In fact if the aluminum were pure, there would be no problem. The potential for trouble comes from whatever other alloys are in the roofing material. To be on the safe side, you could use an insulative material—a layer of roofing felt, neoprene strips or even caulking— between the galvanized snow guards and the aluminum roofing, though this introduces a maintenance headache. Choose a fastener that's compatible with one metal, aluminum lag screws for instance, and then use neoprene washers to isolate them from the other metal.

But the good news is you don't have to bother with all of this. You can get cast-aluminum snow guards from Zaleski Snow Guards and Roofing (11 Alden, New Britain, Conn. 06503).

COATING REMOVAL FROM A METAL ROOF

For five years now I've owned a 125-year-old house and have been puzzling over the problem of its original standing-seam terne roof. It is basically in sound condition, with only a few puncture holes from a hailstorm—no major holes or rust. I would like to restore it, yet the local roofers we have asked say to rip it off because they don't know how to deal with the specific restoration problems.

The roof is covered with a mostly intact layer of aluminum/asphalt paint, applied just before we bought the house. Because of the galvanic reaction, I need to remove this coating, but I don't know the best way. Would sandblasting do the trick? Also what surface protection should I replace it with?

—*Sondra Bromka, Marcellus, N. Y.*

Matt Holmstrom, a remodeling contractor in Bedford County, Va., and a specialist in standing-seam roofs, replies: While 125 years is not extraordinary, it is reasonable for terne, and your roof may be nearing the end of its lifespan. So if I looked at your roof, I might tell you to rip it off, too, if not because of the age, then because of the coating.

I see more roofs that have failed from improper coatings and sealants than from anything else. Asphalt cements and coatings and aluminum paints will ruin terne, the first by trapping water against the terne, the second by galvanic reaction. Since the coating on your roof has been on for more than five years, it may be too late to save the roof. You'll have to decide what to do based on the condition of the metal.

I would not try to strip the coating off by sandblasting. Roof metal is pretty thin. You should try removing it with a high-pressure washer, like those used for stripping paint and cleaning the exteriors of buildings. You should be able to rent one through a good paint store or a tool-rental outfit.

Take the hose up on the roof and blast the whole thing a couple of times from close range. Be careful, though; the water pressure could knock you right off the roof. You will be able to tell right away how successful you are going to be—either a lot of the coating will come off or not much of anything will. Wait a couple of weeks, or even a couple of months, to let the weather loosen up some more of the coating, then repeat the procedure.

If you successfully strip the aluminum paint, finish by wire brushing to remove the loose particles. Then seal any holes in the roof. I recommend using a paintable silicone caulk for small holes. Paint the whole thing with a good rust-preventive paint, like Rustoleum or Benjamin Moore's Retardo. This can be either the primer for a good quality tinner's paint, or you can simply apply a second coat as the finish.

If the aluminum/asphalt coating doesn't come off well, you may still want to apply the rust-preventive paint to improve the appearance of the roof, temporarily. But plan on replacing the roof in the near future.

RECESSED ENTRY OVER FLOOR TRUSSES

I plan to build a home with a recessed entry as shown on the drawing below. This will be a single-story house, with a basement. Floor trusses span the full 32-ft. width of the house. The outdoor-entry floor will be the same floor as the rest of the house, so there is the potential for water to leak through the floor into the basement unless the floor is sealed properly. How should I handle this?

—B. J. Raterman, Falmouth, Va.

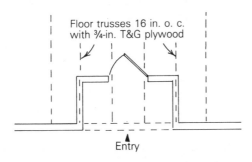

Floor trusses 16 in. o. c.
with ¾-in. T&G plywood

Entry

Tom Reardon, an architect in Northborough, Mass., replies: As you say, there is a very real potential for water leaking into your basement below the entry, and I'm afraid it would be difficult to seal the floor and create a watertight entry. With the entry floor a continuation of the first-floor joists, there will be no slope to the entry deck to shed water away from the house. There may in fact be a slope *toward* the entry door because of deflection in the floor joists.

Also, the ¾-in. T&G plywood subfloor that runs continuously across the floor joists would have to be waterproofed at the recessed entry with a rubber membrane, metal pan flashing or fiberglass—any of which would also need to be covered by a durable exterior decking such as redwood or fir planks. These additional layers would raise the entry floor higher than the door threshold.

For these reasons, I would not recommend recessing the entry as you suggest. You could either eliminate the recessed entry, and perhaps add a small roof projection to shelter the front steps, or you could have the foundation jog in around the recessed entry and drop the entry floor down, treating it separately from the house floor. You could then build the recessed entry with a wooden deck and shallow crawl space below, or you could pour a concrete slab. Either way, you will eliminate the potential for water coming in at the entry.

READERS REPLY
I believe there is another, more appropriate solution to B. J. Raterman's recessed entry in a house with floor trusses. Frame the recessed entry like a stairwell opening, with a doubled floor truss under the side walls. Place a 2x10 or 2x12 header under the recessed entry wall, then frame the entry floor with 2x6 or 2x8 joists. Set the top of the joists at the inboard end an inch or two below the interior floor, and slope the joists about ⅛ in. per ft. to the foundation wall. The foundation wall may need to be stepped down for the recessed framing, depending on whether the floor trusses bear on the top or bottom chord. This solution makes a more secure, and simpler, flashing arrangement than Mr. Raterman's original proposal. Assuming that Mr. Raterman is willing to use a little extra care in this area, I see no reason why he must give up a simple foundation and some basement space to have his recessed entry.

—Gary J. D. Gingras, Plymouth, Mich.

Another simple solution would be to shim the top of each wood truss with a 2x4 laid flat, except in the area of the recessed entry. After he decked the house and entry, Mr. Raterman would automatically have a 1½-in. depressed entry, which he could pitch away from the house and finish up as suggested by Mr. Reardon. Utility grade 2x4s would be cost-effective in this application as the wood trusses would be engineered to carry all loading.

—D. Steve Serda, El Paso, Tex.

HEAT BUILDUP BEHIND STORM DOOR

One of my clients has an all-glass storm door in front of his solid-oak door. The door faces south so you can imagine the heat buildup. The storm door is like a solar collector. It has damaged the finish of the oak door to say the least. Can something be placed on the glass to protect the oak door without reducing visibility through the storm door?

—David T. Gimenthal, North Salem, N. Y.

Geof Cahoon, glazing consultant in Boulder, Colo., replies: The problem here is not so much what can be done to the storm door, but why you need a storm door facing south in the first place. Normally (and this, of course, is subject to the house's particular site) a storm door isn't really needed on the warmest side of the house. This situation is an excellent test to determine how fast you can explode an oak door. Removing the storm door and restoring the finish might be the simplest solution.

If you really need a south-facing storm door, however, here are some approaches you might take. Some sort of traditional bronze or grey shading film applied directly to the glass would help, although to do any good at all visibility would definitely be affected. You are dealing here with two problems. The first is ultraviolet ('UV) light and its damaging effect on the finish. The second is heat buildup in the space between the storm door and the oak entry door. This is caused not only by high-energy light wavelengths (UV) but also by the visible and infrared portions of the light spectrum.

To deal with both problems, more sophisticated glazing solutions are also possible. For example, ¼-in. laminated glass (two sheets of glass with clear plastic film between them) will filter out almost all UV light although the heat buildup would still remain a problem. To tackle both UV light and heat buildup you might consider using a storm door with insulating glass (two sheets with an airspace between them).

With this solution you might include the use of a clear glazing product such as Heat Mirror 55, which would filter out approximately 80% of the sun's UV radiation and reduce visible light transmission, eliminating much of the heat buildup. Heat Mirror (Southwall Technology, 1029 Corporation Way, Palo Alto, Calif. 94303) is a clear plastic film and must be installed inside a sealed insulating-glass unit, so the "storm door" could become rather bulky. In fact, you would actually be using what most contractors would consider a glass entry door in front of your oak entry door.

Finally keep in mind that two panes of regular insulating glass will transmit only 67% of the UV, while using bronze low-E glass in an insulating-glass storm door unit would cut the UV down to 31%. However, using some of the "low-E" glass products would not help a great deal in this situation because of their limited effectiveness against UV transmission.

Another solution would be to create some ventilation from the interior of the house into the enclosed space between the entry door and the storm door. I've seen this done in the past with a combination of decorative louvers or small operable windows and "flap valves" that prevent reverse thermo-siphoning at night. This would lower the temperature in the space and leave you with only the problem of direct southern exposure through the glass of the storm door. This can be dealt with through the use of any of the urethane exterior-grade finishes with UV-resistant filler and pigments.

SIZING SUPPLY LINES

We are remodeling the ground-floor bathroom in our four-story, two-family brownstone. We will be replacing the water-supply pipes and are having trouble figuring out what size pipes to use. Can you tell us what to use for best performance or suggest some reliable references?

—Mark Goldfield, Brooklyn, N. Y.

California plumber George Skaates replies: To size your water system properly, I suggest that you buy a copy of the 1988 Uniform Plumbing Code (UPC) from the International Association of Plumbing and Mechanical Officals, (20001 S. Walnut Dr., Walnut, Calif. 91789) or get one from the library.

Before you can use the book, though, you should collect some information. First, make a list of all your plumbing fixtures, including all hose bibbs and lawn-sprinkler heads. Next, figure out the developed length of pipes (the actual total length of pipes) between your meter and your farthest outlet. Calculate the difference in elevation between your meter and the highest plumbing fixture, and determine the incoming water pressure from your meter.

After you have this information, refer to Chapter 10, "Water Distribution," in the UPC (pp. 73-84). Table 10-1 will help you with your fixture count, and table 10-2 will give you the correct pipe size. Be sure that you refer to the section that reflects the incoming pressure at your meter. Start at the most remote outlet and compute your pipe size using table 10-2.

INDEX